John Thomas Wheeler

Professional Achievement for Engineers and Scientists

Books by Tyler G. Hicks

PLANT ENGINEER'S EASY PROBLEM SOLVER

PUMP SELECTION AND APPLICATION

PUMP OPERATION AND MAINTENANCE

SUCCESSFUL TECHNICAL WRITING

WRITING FOR ENGINEERING AND SCIENCE

INDUSTRIAL HYDRAULICS (*with John Pippenger*)

PROFESSIONAL ACHIEVEMENT FOR ENGINEERS AND SCIENTISTS

Professional Achievement for Engineers and Scientists

How to Earn More Money and Greater Success in Engineering and Science

TYLER G. HICKS

Mechanical Engineer; Associate Member, American Society of Mechanical Engineers and United States Naval Institute

Member, The International Oceanographic Foundation

Editor, Proceedings of the Oil and Gas Power Division, *ASME*

McGRAW-HILL BOOK COMPANY

New York San Francisco Toronto London

PROFESSIONAL ACHIEVEMENT FOR ENGINEERS
AND SCIENTISTS

 Library of Congress Catalog Card Number 63-13934

28739

II

This book is dedicated with respect and understanding to my fellow engineers and scientists who are striving to achieve more in the expanding world of technology and science. May your contributions be significant and your efforts rewarded

Preface

This book is written for and is useful to every graduate engineer and scientist—particularly those out of school five years or more. Why five years? Because much observation and study led me to conclude that thousands of engineers and scientists become dissatisfied with their careers about five years after they obtain their degrees—graduate or undergraduate. Sure, some men begin to complain sooner—others are satisfied for more than five years. But the five-year interval is reasonably accurate. Think about your own career—and the careers of your friends. You'll see this statement is approximately true.

Now what will this book do for you, Mr. Engineer or Mr. Scientist? *It will help you earn and achieve more in your profession*. How? In many ways. It will help you see your place in modern technology and science. This book shows you how to analyze your current job and earnings—and your future prospects. You are shown how to redefine your present goals, or choose new ones.

Should you get into management? This is a major question every engineer and scientist faces at some time during his career. This book *shows* you what to expect from a management career versus a technological career. Using the guideposts cited, you can make the big decision. Once your decision is made, you can go on to an outstanding career in the field of your choice. You are shown how to train yourself in the skills and techniques you need to increase your income and to lead a full and rewarding life in either technology or science.

And what are these skills and techniques? The table of contents lists them. But here are a few—how to earn money as a full- or part-time consultant; serving and using your professional society to best advantage; thinking creatively to increase your income; making business travel pay; organizing your workday for best efficiency; improving your knowledge of your field; making human relations pay off; how to write, speak, and listen better. Will learning these skills and techniques make you a better engineer or scientist? Definitely yes. Will your paycheck be larger? Definitely yes—you can't miss if you apply yourself.

Colleges and universities do an excellent job of training engineers and scientists for the professional aspects of their careers. But in the crowded academic curriculum there is little time to teach many of the important nontechnical aspects of engineering and science. This book fills some of the gap. Planned and written for the mature engineer and scientist, the book is hard-hitting and practical. It shows you *exactly* what you can do with your career, and how to achieve your financial and creative goals. The book analyzes *your* potential returns in terms of your expended effort. You are guided to actual, achievable objectives that will increase your income and your contributions in a given field. Numerous examples from the lives of outstanding engineers and scientists show you how to achieve more in your life—both in the professional and personal phases. Not only will you earn more for your efforts, you will also obtain a deeper satisfaction from your career than ever before.

Think, for a moment, how just a few of the following hints might change your life—ten ways to develop a spare-time income with little or no capital; five ways to economize on your talking time; thirteen simple checks for a professional engineer's license eligibility; twelve keys to speeding your paper work; fourteen problems to avoid in becoming a full- or part-time consultant; ten ways to read for profit; eight steps to releasing your creative powers to build your income; seven ways to collect data to solve a problem; nine steps for effective planning; thirty check questions to help you decide if you should change jobs. These are only a few of the thousands of practical techniques you will find in this book.

Can engineering students use this book? Most certainly yes. But a word of caution to young students—you won't find an equation, theorem, or law in this book. For the guideposts to professional achievement cannot be stated in terms of an easily remembered integral or differential. Instead, the skills that lead to success in any area of engineering or science must be learned by trial and error in the daily practice of a profession. Students using this book will reduce the number of tries they must make before they master a given skill. The book will also help eliminate many painful errors during their careers. So no student will lose, if he carefully applies the techniques given here.

This is a needed book—needed by all engineers and scientists and the professions they represent. More than a decade of close contact with thousands of engineers and scientists in all fields has convinced me that greater recognition and stronger professionalism will not come from idle wishes and hopes. The only way engineers and scientists can achieve a higher place in society and the community is by working for this goal every moment of their careers. This book, then, shows all engineers and scientists in every field how they can begin to achieve greater success, earnings, and happiness in their profession. Use this book now—it can become the first milestone in a fuller, happier life for yourself and your family.

Tyler G. Hicks

integral or differential. Instead, the skills that lead to success in any area of engineering or science must be learned by trial and error in the daily practice of a profession. Students using this book will reduce the number of errors they must make before they master a given skill. The book will also help eliminate many painful errors during their careers. So no student will lose if he carefully applies the techniques given here.

This is a needed book—needed by all engineers and scientists and the students of these professions. More than a decade of close contact with thousands of engineers and scientists in all fields has convinced me that greater recognition and stronger professional status will not come from idle wishes and hopes. The only way engineers and scientists can achieve a higher place in society and the community is by working for this goal every moment of their careers. This book shows them—engineers and scientists in every field—how they can begin to achieve greater success, earnings, and happiness in their professions. Use this book now—today—to start the journey to a fuller, happier life for yourself and your family.

Tyler G. Hicks

Acknowledgments

While writing this book I consulted hundreds of engineers, scientists, managers, and others familiar with the traits, working habits, and problems of today's technical and scientific personnel. For five years prior to the start of the writing I collected every book, article, survey, and opinion poll related to the practicing engineer and scientist that I could find. To these people, and to the authors of hundreds of written pieces, as well as their publishers, I am deeply indebted. Their breadth of experience, generosity in granting permission to quote important findings, and suggestions for subjects to cover, made the writing of this book a most rewarding experience. While I have tried to acknowledge each contribution in the chapter in which it appears, I would like to give additional credit here to those whose names do not appear in a chapter for lack of space, or for other reasons. In this way I hope to gratefully acknowledge the help and cooperation received from many sources. In the order of their appearance in the book these are:

Joseph C. McCabe, editor and publisher, *Combustion;* Elmer Tangerman, editor, *Product Engineering;* James E. Hughes, vice-president, Western Supply Company; F. E. Satterthwaite, director, Statistical Engineering Institute; James M. Black, Princeton, N.J., whose leadership tests are so useful; Alvin Schwartz, director of communications, Opinion Research Corporation, whose surveys of engineers and scientists provide valuable insights into the minds of today's technical and scientific personnel; Frank Coss, vice-president, Deutsch & Shea, Inc.; Gerald L. Farrar, engineering editor,

The Oil and Gas Journal; Auren Uris, whose quotations in Chapters 3 and 31 are reprinted by special permission from *Dun's Review and Modern Industry,* copyright 1963 by Dun & Bradstreet Publications Corp.; Steve Elonka, associate editor, *Power;* R. P. Kroon, Burleigh B. Gardner, Professor Oliver J. Sizelove, Department of Management Engineering, Newark College of Engineering, and Eugene Raudsepp, director of psychological research, Deutsch & Shea, Inc., whose studies, surveys, and writings led to the list of qualifications of managers in this book; W. E. Dewey, knowledgeable author on more effective use of time in business life; William S. Schaill, for his hints on building reading speed, reprinted from *This Week Magazine,* copyright 1961 by the United Newspapers Magazine Corporation; McGraw-Hill Publishing Company, Inc., The Reading Laboratory, Inc., and *Executive's Digest,* for general hints on increasing reading speed; Winston Brebner, Cambridge Associates, Inc., whose excellent *Executive's Digest* served as a source for many stimulating ideas about the place of the engineer and scientist in today's business and technical community; E. D. Gray-Donald, Deputy General Secretary, The Engineering Institute of Canada, and Elsie Murray, executive secretary, Engineers' Council for Professional Development; William N. Beadle, Byron Jackson Division, Borg-Warner Corporation; Lester Bittel, editor, *Factory;* Alex F. Osborn, L.H.D., of Batten, Barton, Durstine & Osborn, Inc.; C. D. Tuska, director of RCA Patent Operations; Charles Scribner's Sons, publishers of Alex Osborn's *Applied Imagination;* G. G. Hawley, executive editor, Reinhold Publishing Corporation; Charles S. Whiting, Market Planning Corporation (an affiliate of McCann-Erickson, Inc.), author of *Creative Thinking;* M. H. Johnson, Editorial Department, Stanford University Press; Philip H. Abelson, editor, *Science,* and Dr. F. Reif, Department of Physics, University of California, Berkeley, Calif., for permission to reprint from *Science* a portion of one of Dr. Reif's articles; Tom Johnson, associate editor, *American Machinist/Metalworking Manufacturing; Carnegie Alumnus;* J. J. Jacklitsch, editor, *Mechanical Engineering; Space/Aeronautics;* Kenneth Kramer, managing editor, *Business Week;* Cecil H. Chilton, editor, *Chemical Engineering;* Robert B. Buck, president, Executive Counselling Service, Inc.; James A. McCole, Jr., News Department, Dow Jones & Company, Inc., and *The Wall Street Journal;* Paul Gaynor, Gaynor & Ducas, Inc.; Silas B. Ragsdale, editorial director, *Petroleum Refiner* and *Hydrocarbon Processing,* for the quotation by Major Bronson Foster, copyright 1957 by Gulf Publishing Company, Houston, Texas; Edward L. Anthony, Small Business Administration, Washington, D.C.; M. M. Matthews, managing editor, Westinghouse *Engineer; New York Herald Tribune;* Elizabeth H. Knox, permissions editor, *Harvard Business Review; Petroleum Refiner,* for permission to quote short hints on leadership, copyright 1956 by Gulf Publishing Company, Houston, Texas; Kenneth Schwartz, senior editor, *Dun's Review and Modern Industry,* for the Clarence B. Randall quotation, reprinted by special permission from *Dun's Review and Modern Industry,* June 1959, copyright 1963 by Dun & Bradstreet Publications Corp.; Harold F. Smiddy, vice-president, General Electric Company; Kent McKamy, editor, *Business Management;* John R. Wells, associate editor, *The Saturday Evening Post,* for the details about Donald Doughty, Ph.D., in Chapter 18, reprinted by special permission from *The Saturday Evening Post,* copyright 1962 by The Curtis Publishing Company; B. H. Angert, business manager, Adult Education Association of the U.S.A., and *Adult Leadership;* James B. O'Connell, managing editor, *Think Magazine,* for the Ernest Dale quotation in Chapter 19, reprinted by permission of *Think Magazine,* copyright 1961 by International Business Machines Corporation; Kent McKamy, editor, *Management Methods;* L. N. Rowley, editor and publisher of *Power,* whose ideas on technical-paper presentation are included in Chapter 20; Colin Carmichael, editor, *Machine Design;* John Constance, P.E., well-known licensing consultant whose useful hints on professional licenses appear in Chapter 24; Dr. D. H. Ewing, whose quotation in Chapter 25 is reprinted by special permission from *Dun's Review and Modern Industry,* copyright 1963 by Dun & Bradstreet Publications Corp.; John Whiteman, manager, External Communications, Public Relations Department, Corn Products Company; William D. Falcon, editor, *The Executive;* Robert L. Calio, managing editor, *Administrative Management;* Howard Simons, whose quotations in Chapter 28 are reprinted by permission from *Think Magazine,* copyright 1961 by International Business Machines Corporation; Stuart Chase, whose quotation in Chapter 29 is reprinted by permission from *Think Magazine,* copyright 1960 by International Business Machines Corporation; Brooke Alexander, assistant to the publisher, *Fortune*

Magazine, for the quotations on small business in Chapter 29, copyright 1954, 1955, 1956, and 1957 by Time, Inc.; Sam T. Griene, *Supervision* Magazine; A. C. Spanger, *Advanced Management;* Peter J. Celliers, author of "the six keys to a good memory" given in Chapter 31; Edward Oxford, *Telephone Review;* H. F. Merrill, *Supervisory Management,* and The American Management Association, Inc., for permission to quote J. M. Black in Chapter 31; M. E. McNeill, *Public Relations Journal;* Harold J. Swarl, Printers' Ink Publishing Corp.

To these, and to many others, the author offers sincerest thanks for their assistance and advice. Lastly, my eldest son, Greg, deserves thanks for his help in preparing the index.

Contents

Part 3. Broaden Your Professional Horizons

PART 1

Develop Your Personal Skills

1

Choose Your Goals Today

> No age or time of life, no position or circumstance has a monopoly on success. Any age is the right age to start doing something.
>
> WYNN JOHNSON

Today as an engineer or scientist you can't lose. Opportunities for expanding your contributions to your field are growing so fast you could never, in one lifetime, take advantage of them all. You have a wider choice of technical and scientific career specialties than any man ever imagined. Your earnings are at the highest level in history. More firms than ever before want to put you on the payroll—and keep you there. Project managers and company executives actively seek greater creativity on the part of every engineer and scientist.

Do you seek to make an original, important contribution to your speciality? You can—if you try. Multimillion-dollar labs are available to you, if you choose the right company to work for. You'd rather be on your own as a consultant? Fine. More consultants have more clients than ever before. You'd rather get into management or run a project or a group? Go right ahead. Some 40 per cent of America's executives have a technical background. Looking for greater happiness for yourself and family? It's yours, once you decide what will bring this happiness to you and yours—a different job, a home in another locality, work in civic affairs.

Look Around—Count Your Advantages

Take a look at yourself, your career. If you're with a large firm, chances are good you work in modern quarters with excellent lighting, up-to-date furnishings, understanding supervisors. There's a good technical or scientific library in the office or plant. You can read the latest journals, magazines, or books. The boss doesn't frown—he wants you to keep up.

Working conditions are good; methods are modern. You have access to top-notch dictation equipment. Secretaries and technicians help you with routine chores.

Want to attend an engineering or scientific conference or meeting out of town? Tell the boss; then call the transportation department. Your tickets, paid in full, are on your desk the next day. You pack and go—for a day, a week. While at the meeting you learn the latest in your specialty and meet fellow engineers and scientists in your field. Experiences, data, and ideas are swapped—you return with a fresh viewpoint.

Your raises are regular. Few raises are ever as large as you hoped. But raises do come through. One day you sit back and count your advantages. You're surprised to learn that you're doing reasonably well. For the most part you're as happy as a man can expect to be.

But what about the engineer or scientist with the medium-sized or small firm? Does he have these advantages? Yes, in most modern medium-sized and small firms. Some smaller firms give their engineers and scientists even more freedom than larger ones.

Universities, research institutes, and government organizations? Most engineers and scientists in these activities have as much freedom as those in industry, or even more. While universities may not have as liberal travel budgets as industrial firms, the greater time available for study and personal development offsets this slight disadvantage. Also, the opportunities for pure and applied research may be greater.

But You Still Want to Do More?

Fine! You wouldn't be worth much to any employer, or yourself, if you were satisfied with every phase of your career. For as Bill Hindenlang, associate editor of *Combustion* magazine recently said:

> Young engineers with enthusiasm and ideas might profitably lay aside for a moment their pension plans, major medical benefits, paid vacations and holidays and look about for the companies who are moving ahead—who are risking real capital on the wheel of fortune. Older engineers who find themselves deep in a security-plowed rut may ponder the difference between a rut and a grave. In a rut one can still see the sky and climb toward it.

What is this sky? It is a goal or goals toward which you can work. Every engineer and scientist had a goal early in his career. This goal was his degree—either undergraduate or graduate. Without this goal few engineers or scientists would have ever finished their formal education. The goal of the degree helps all engineering and science majors to channel their energies into learning the many techniques and facts needed in their careers.

But what happens after the degree is won? Many engineers and scientists, failing to understand the importance of goals, limp along in the soft, security-lined rut. Their biggest goal in life is to catch the 5:15 early enough to get a good seat. Or their daily goal may be the number one position in the stream of cars leaving the parking lot. Being first on the northbound freeway seems to give greater satisfaction than any other achievement.

You *can* achieve better goals in life. How? By applying the techniques given in this book. But before you can achieve more you must choose your goals. Here's how you can start today to decide which goals are worth your effort.

Answer Six Questions

Analyze your present thinking by answering the following six questions. Don't be discouraged if it takes you several hours, or days, to answer these questions. They were designed to make you probe deeply into your personality. *Write the answer to each question on a separate sheet of paper.*

1. What is your ultimate career goal?
2. How will this goal help you achieve more in life?
3. Where (industry, government, education, etc.) can you best fulfill this goal?
4. Which specific organization will best help you fulfill your goal?

5. When can you reasonably expect to fulfill this goal?
6. Why have you chosen this career goal?

Spend some time seriously considering these questions. The first question—your ultimate career goal—will be the most difficult to answer. For no matter how high a position you may aspire to, the process of accurately defining it can be frightening. Writing out your answer to each question will put your goal in front of you with a finality you never thought possible. You will be better able to judge whether your goal is right for you.

But what if you, like many engineers and scientists, can't name a specific ultimate career goal? What should you do? Cast the question aside? No; you must, if you wish to achieve and earn more, answer these questions. Some men can get by with a fuzzy idea of what they ultimately hope to achieve. But most engineers and scientists, trained to think in terms of the specific solution to a given problem, are lost unless they can put their goals on paper. Once the goal is written out, most engineers and scientists can begin to take steps to achieve the goal.

You still cannot answer question 1? Then try these amplifying questions: Just what do you want out of life—money, position, power, leisure? What position in your chosen field will best allow you to achieve any or all of these goals? Would you be happy in that position? Can you achieve that position in your present organization? Is there any additional training or study needed to best fit you for this position? Objective answers to these questions will almost certainly permit you to answer question 1.

Amplify Your Answers

You now have six sheets of paper, each with an answer at the top. Expand each answer so it fully discusses your goal. Thus let's say your answer to question 1 was *chief design engineer in the field of solid propellant carrier vehicles*. Write out a description of the responsibilities, skills, and rewards of such a goal. Be specific —name the types of knowledge and training required, the number and types of people you would supervise, the probable annual salary, the related duties—travel, preparation of reports and technical papers, selection of key personnel, preparation of budgets, management conferences, etc. Prepare this comprehensive word picture of your goal, using your knowledge of the duties of the chief design engineer in your own or other organizations. If you don't know a person holding such a position, do some research in your technical library and among your friends in the profession. You can learn much about your proposed goal by using these and similar simple procedures.

Prepare similar write-ups for the answers to the other five questions. Answering all six questions will force you to face up to your goal with more objectivity than you may have ever mustered before. You will be able to evaluate your goal in the privacy of your mind without having to commit yourself to its attainment.

Evaluate Your Goal

Examine your write-ups. Ask yourself:

1. Is this goal in accord with my temperament?
2. Am I willing to live with the stresses this goal will enforce on me?
3. If I aspire to more leisure, do I know what I'll do with the extra time when it is available?
4. If my goal is a large sum of money, will I be able to put the money to good use? (Money of itself is useless unless put to work.)
5. Will the probable annual income of my goal satisfy the needs of myself and my family?
6. Does this goal provide me an opportunity to make a lasting contribution to my chosen field?
7. Can I pursue this goal in an area of the country or the world which also meets my other interests? (Thus a ski enthusiast might be unhappy if his career goal were resident chief engineer of the Panama Canal.)
8. Does achieving this goal require me to ignore any moral or religious beliefs I may have?
9. Will this goal contribute to the betterment of mankind?
10. Is this goal truly what I want in life? Or have I mistaken this goal for just one step on the way toward another, greater goal?
11. Can this goal be achieved without a major change in my present way of living?

12. Does this goal offer a chance to continue my career, on a part-time basis, after my formal retirement from my present or any future organization I may serve?

Don't delude yourself into thinking that your goal need be chief engineer, chief scientist, president, or some similar high station. If your true goal is senior engineer, staff scientist, traffic manager, or some similar position, fine. Not all of us want to be, or can be, top man in the organization. You may be able to make a more important contribution as a staff scientist than you ever could as chief scientist in charge of a research laboratory. Remember this general rule: *The higher your position in any organization, the less technical or scientific and the more managerial will your duties be.* So you must decide in which general area—technology or management—you most sincerely desire to spend the remainder of your career.

Does choosing a goal today limit your future freedom of choice? No—you will be constantly reevaluating your goal as time goes by. But having a goal to pursue will direct your efforts to a much greater degree than you may now think possible. Just having a specific goal helps alert you to the importance of acquiring new skills and techniques. Then if future experience shows that a new goal would be more suitable, you will be better able to recognize and achieve it. The very act of choosing a goal is excellent training for every engineer and scientist because few of us see our career and its possible contributions to mankind and ourselves as clearly as we might.

Specify Lifetime Goals

So far in this chapter we've discussed how you can more clearly define your career goals. Now let's examine typical lifetime goals you might set for yourself. While not every goal will apply to you, those cited will stimulate your creative thinking and lead you to more worthwhile achievements throughout every phase of your life.

Divide your lifetime goals into three categories: (1) professional, (2) public, and (3) personal. Let's examine each category to see how you as an engineer or scientist can fulfill it.

Professional goals. Probably the most important professional goal you could choose is that of making a useful contribution to your field. Thus, you might: (1) develop a new or improved technique for manufacture, operation, analysis, etc., in your field; (2) investigate and state a new theory in your area of technology or science; (3) prepare a written standard that can be adopted to help make practices more uniform in a given area of technology; (4) spread news of your field by preparing and circulating publicity information on various levels for the general public, your associates, people in other fields, etc.; (5) educate people about your field by giving courses, speaking at schools and colleges, holding seminars, and writing about the field in suitable publications.

Achieving some or all of these goals can lead to other accomplishments like greater prestige for you in your field and organization; more satisfaction from your work and profession; growth of your professional knowledge and skill; greater earnings; more leisure; increased responsibility; and freedom of action. Though all of us may desire larger earnings, greater freedom, and more leisure, it is not until we recognize that these are best attained by making a genuine contribution that we truly begin to move ahead in our profession.

Public goals are important. Today more engineers and scientists are learning that they can make useful contributions to their communities by public service of some kind. While lawyers and medical men have long been recognized as important in community activities, engineers and scientists are just coming into their own. The growing complexity of our cities and towns makes the engineer or scientist a welcome addition to every public group. So you now have more opportunities than ever before to take an active part in democracy. Let's see how you might devote your time and energies.

Public areas in which you might serve include: (1) education—your local school board, high school student advisory council, parent-teacher association, benefit drives, etc.; (2) local-level politics—town council or board, elected part-time offices (mayor, sheriff, etc.), political parties of your choice, and similar activities; (3) youth groups—Boy Scouts, Little League, Girl Scouts, etc.—in youth-group service you can help your own and others' children prepare themselves for a better life; (4) religious activities in the belief of your choice where you can help in fund drives, services, youth or adult clubs, building and grounds maintenance, etc.

Public activities of any kind will do much to broaden your possibilities for important achievements. You will meet new people, face and solve unique problems, aid in your area's continued growth, and contribute your energies toward the betterment of mankind in activities where your talents can be used. Don't spurn public duties because they contribute nothing toward a larger income. You will find that the skills and friends you acquire in public service will often assist you in many ways in your professional and personal goals.

Personal goals pay off. Having goals in the areas of your profession and public service will give greater meaning to the hours you spend away from home. But to round out your life so you become a fully effective individual you must have goals that build your personal life to its fullest potential. These goals lie in the general areas of your family, home, hobbies, and education. Let's look at these in reverse order because this is one way of seeing the interplay of these areas.

Education, properly pursued, is a lifelong task. As you'll see later in this book, you'll need every bit of energy and enthusiasm you can muster just to keep up with developments in your own field. But don't evaluate education only in terms of technology and science. There are thousands of other fields you can learn about during your life. These fields may relate to your hobby—like photography, boating, skiing, fishing, golf, hunting, etc. Or they may relate to human relations skills, politics, youth groups, and many others. Set a goal to learn at least one new subject each year of your life. Even a simple goal like learning as much as you can about automotive mechanics can broaden your horizons, increase your skill, and give you many hours of pleasure. So stop for a moment and examine your interests and needs in the educational area. Choose two subjects—one related to your profession and the other to your outside interests. Resolve now to master each subject in a specified time. See Chapters 25 and 29 for useful suggestions.

Hobbies, like education, can renew your interest in life and expand your personality. A creative hobby can refresh your mind so you work more productively. Most hobbies can be pursued inexpensively. Many allow you to relive your childhood interests. Some hobbies, when shared with your children, bring you closer to your growing sons and daughters. Going back to a former hobby when you're in your thirties or forties can recall your youth with sharp vividness. Sharing this hobby with your kids will give them a better insight into their father than all the lectures you could ever give. So revive an old hobby or develop a new one—model trains or boats, painting, music, shells, guns—what have you. The hobby may be a better tonic than any medical man might prescribe. And it may lead you to a closer and more fruitful family life than you've ever enjoyed before. Set a new goal—an unusual goal—a hobby just for the fun of it.

Many a professional man's home is his hobby. He tinkers and putters, improving the livability and value of his home while relaxing at the same time. Choosing the location and furnishings for a comfortable home can solidify family interests, build cooperation and respect. So set a goal today to make your home the center of your interests and after-hours efforts. You and your family will benefit immeasurably.

Don't overlook your family in the new goals you develop for yourself. Try to visualize achievements that will benefit your wife and your children. Typical family goals that will be useful to all are travel to historic areas during your vacations, educational hobbies and games, attendance at cultural exhibits or performances By making your family a part of your lifetime goals you will broaden their lives while drawing them closer to you. Lastly, don't overlook the many opportunities to let your family share in your profession by visiting your place of work, accompanying you on business trips, or attending parties and other functions given by your organization.

Greater Achievements Can Be Yours

The remainder of this book is devoted to showing how you, as an engineer or scientist in any field, can achieve more in life. Not every hint and technique given will suit you—but most will. Even if a few of the suggestions don't seem right for your life, they will stimulate your mind, helping you to come up with better ideas. Try to apply as many of the hints as you can. For if you do, you're certain to have a fuller life and greater achievement in your profession.

Hundreds of engineers and scientists who have followed the suggestions given in this book have benefited in many ways. Increased earnings, greater achievements, and a sense of fulfillment are a few of the benefits derived. You, too, can share in their achievements. What's more, following the hints given will *not* harm you—every hint given is a positive one that will bring you one step closer to greater professional success in any field of technology or science.

HELPFUL READING

Berelson, Bernard, *Graduate Education in the United States,* McGraw-Hill Book Company, Inc., New York, 1960.

Davis, Keith, *Human Relations at Work,* McGraw-Hill Book Company, Inc., New York, 1962.

Haldone, Bernard, *How to Make a Habit of Success,* Prentice-Hall, Inc., Englewood Cliffs, N.J., 1960.

Mason, Joseph G., *How to Be a More Creative Executive,* McGraw-Hill Book Company, Inc., New York, 1960.

Schwartz, David Joseph, *The Magic of Thinking Big,* Prentice-Hall, Inc., Englewood Cliffs, N.J., 1959.

Soule, George, *Time for Living,* The Viking Press, New York, 1955.

Uris, Auren, *The Efficient Executive,* McGraw-Hill Book Company, Inc., New York, 1957.

2

Should You Get into Management?

> Labor can do nothing without capital, capital nothing without labor, and neither labor nor capital can do anything without the guiding genius of management. . . .
>
> W. L. MACKENZIE KING

Today more American engineers and scientists are moving into management positions than ever before in history. Why? There are many reasons. Talk to a number of these transplanted engineers and scientists, and you'll hear certain reasons repeated. Usually these are: greater opportunities for advancement; higher salary with chances for profit sharing; more challenging work; broader contacts with people; wider chance for personal expression; better status in the community. Of course there are other reasons. But most of these will not differ much from the above six.

Should You Switch?

Listen to the engineers and scientists at technical meetings, refresher courses, and on the job. Time after time you'll hear a man say, "I'm thinking of changing to managerial work. But I haven't made up my mind yet."

Some engineers and scientists spend their lives thinking about making the switch and never do. Others think through the problem and make the change. Some of these engineers and scientists are happy and successful in their new work. But other men find they prefer engineering or science to management. As one engineer put it, "I'm happier in R & D than I ever could be in any other work. At heart I'm an engineer, and that's what I'll stay."

You, and only you, can answer the question: [1] Should I switch from engineering to management? And to answer this question accurately you must know yourself. You must know what you want in life, where you want to go, and how much effort you're willing to put into your career.

But suppose you do change to management. What new problems will you meet? What new skills must you develop? How will your rewards compare with those for a job well-done in engineering or science? Let's see.

Specific to general. In engineering or science you work with specifics—weight, length, height, pressure, force. In management you work with generalities—supervision, arbitration, sales, delegation, negotiation. Engineering and science have their neat slide rules, graph paper, handbooks, laboratory tests, and formulas. After a few years you become a specialist in one or two areas. These become comfortable, well-defined boundaries which give you satisfaction and a livelihood and allow you to contribute to technology. You always know where you stand because you can accurately measure results. And you can be as exact as you wish. Your tools are known; your theories can be tested and proved by accepted methods.

Management, too, has its tools. But compared to the precise tools of engineering and science the tools of management are crude. Why? Because the problems of management are affected by so many unpredictable variables—human, economic, and political.

So the first difference you'll find between engineering and management is one of *specifics versus generalities.* In engineering there is almost always an equation you can use or develop. In management there is hardly ever an equation you can apply.

During your first few months in management you'll find yourself trying to plot curves for or tabulate nonmeasurable data. After several attempts you'll give up. And then you may say to yourself, as one new engineering manager did, "Everything here is gray; there's no right answer."

That's just it. There isn't any "right" answer. There's only a workable answer that may be right today and wrong tomorrow. Once you realize this, you'll throw away your graphs and tables. You'll mentally balance yourself like a boxer on his toes, ready to change with external conditions. For your world is no longer neatly bounded by slide rules, formulas, and handbooks. Instead it has no real boundary and its elements are constantly changing.

People, people everywhere. In engineering or science you figured a beam, designed a new machine, or wrote a specification. In management you delegate work, settle disputes, approve expenditures for projects whose outcome is a gamble. But always there are people.

Some engineers and scientists work on a project for weeks and have little or no contact with other people. As part of a management team you can hardly work five minutes without dealing with other people. And there are all kinds of people—happy, angry, disappointed, engaging, healthy, and sick.

You must deal with them in hundreds of ways. You will delegate work, settle disputes, reprimand, explain, answer questions, decide between alternatives. In management you may spend as much as 80 per cent of your time talking with people. And it isn't always easy, particularly when you're trying to convince, sway, or influence people.

Probably the most difficult part of changing from engineering or science to management is learning to cope with the irrational verbal and written demands of people outside your organization. Your first reaction as an engineer or scientist will be to search for a logical, factual answer to complaints or insistent demands of someone who has little knowledge of your business. After a few attempts you'll find that a completely logical and factual answer is seldom satisfactory. Why? Because you're dealing with human beings. You'll push your engineering books to the corner of the bookshelf and replace them with a few books on psychology and human relations. Once you do this, you're well on your way to becoming an educated manager.

Even some of your managerial associates may puzzle you at first. For instead of operating in accordance with equations, theorems, or physical laws, they seem to operate by means of some strange, unseen mental waves. Actually, there's nothing strange about their ways. These experienced managers are constantly analyzing, judging, and evaluating business situations. And they're rapidly altering their opinions, concepts, and theories as they meet new situations.

You, too, must learn to think faster. Many of your quick decisions may be wrong. But instead of fretting, as many engineers and scientists are inclined to do, you must rectify your decision and go on to your next problem. For in dealing with people instead of facts you seldom have time to spend days and days analyzing a situation. People are often hurried; there's always a phone ringing; and someone is impatiently waiting in the lobby to see you.

Your reading will change. Today, as an experienced and skilled engineer or scientist, you probably read a dozen or more technical magazines and journals every month. You study all kinds of good articles—design procedures, maintenance operations, new equipment, etc. You're looking for new ways to do a better engineering or scientific job. Specific procedures, exact details, and proven values are of greatest importance to you. These are data you need to do a top-notch professional job.

Change to management and you'll alter your reading habits. You'll still read the same magazines, plus a few more, but your attitude toward the subjects will change. Instead of trying to understand every step in a process or procedure, you'll look for the results. Your eye will seek out hourly production rates, costs, manpower needs. You'll mark an article for the assistant chief engineer with, "Jim, please check this process; it might save us some money on the new design." And Jim will check the details while you present the idea to upper-level management.

Personnel columns will mean more to you in management than they did in engineering or science. You'll seek out the news of promotions, retirements, and transfers. Because once again, your main concern is with people and their effect on your organization or business.

You'll read different books. Texts on finance, production, sales, speech-making, patents, inventions, human relations, and publicity will nudge your engineering handbooks and texts aside. You'll seek general methods—not specific procedures. You will begin to operate by feel instead of by fixed formulas or rules. The generalist aspects of your character will emerge to take over and push the specialist training of your engineering or scientific career into the background.

You'll make speeches. Management solves many of its problems in meetings, conferences, and panels. You'll be called on to speak. In a company meeting you'll be expected to explain, defend, propose, excuse, or present ideas of all kinds. Much of your future will depend on how well you act at these gatherings. For higher-level management will not only be judging your business decisions—they'll also be evaluating you as an individual. So to succeed you'll have to learn how to talk and think on your feet.

The company meeting is not the place for nervousness, slips of the tongue, or immature self-centeredness. So get to know yourself better. Remember that your engineering or scientific training is a priceless asset. As an ex-engineer or scientist you'll be respected by nonengineering management. They'll look to you for technical opinions and judgments involving production, design, and maintenance. So forget your nervousness; concentrate on what you're saying. But don't try to impress them with your personal intelligence and special education. Slice your opinion to the bone—give results, costs, advantages. Omit long-winded step-by-step procedures. These take up valuable time. Remember: You are now a generalist, not a specialist. Use these hints, and you'll soon see a new you emerging—a valuable and respected you, a real asset on any management team.

You'll also speak in public to audiences of all sizes. And you'll be frightened at first—everyone is. But stick to it. Prepare your speech early; practice long. Use the hints in Chapter 20. Or get yourself several good books on public speaking. If these don't give you what you want, seek professional advice. Today there are speech-training classes in almost every large city.

Your success as a manager will depend, to a large extent, on your ability to motivate other people. So learning how to speak well before others is one of your most valuable tools. Start developing your speaking skill today.

Your thinking will change. As part of the management team you'll think differently. Instead of being confined to a single engineering or scientific project, your viewpoint will be company-wide. Overall profits and losses will be of extreme importance. Also, the financial performance of your own department will be a daily concern. You'll begin to see how your efforts affect the entire organization.

To many ex-engineers and scientists there is nothing as challenging as guiding a group of people along the path to high profits.

Your methods of thinking will change too. You'll find that you must:

- Think of others first.
- Think ahead of your present problems.
- Think of many alternatives.
- Think in terms of selling.
- Think of new-business sources.
- Think while listening to others.
- Think to get to the core of problems quickly.

Of course there will be other ways in which your thinking will change. Much depends on your firm, its products, and your new job. But you must realize now that your thinking must change in many ways if you are to succeed in management.

Don't resist these new thinking patterns when they are introduced to you. Some engineers and scientists resist the novel situations management work forces on them. This causes inefficiency and makes the change from engineering to management a painful process. Remember: Once you decide to change from technical work to managerial tasks you must wholeheartedly accept your new environment. If you do not put all your energy and skill into the new tasks you face, your road will be rough and full of disappointment. So you must learn to think as a manager.

You'll train people. As part of the management team you'll have the problem of training people for new or different jobs. At first your training duties may be no more than instructing a secretary about her job. But as time passes you may be called on to train people for more responsible positions. When faced with this task you'll begin to realize the importance of right thinking, ability to handle people, and good speaking habits.

Training others can be one of the most gratifying aspects of management. It gives you a chance to pass your experience along to others for their ultimate benefit and the benefit of the organization. And the questions and discussions that arise during training sessions will open new vistas of self-knowledge and job skills to you. You will see, more clearly than ever before, that the human factor is the most important single element managers must handle—every day of the year.

Human relations—six handy rules. Many organizations now have clinics in human relations for all new management personnel—particularly ex-engineers and scientists. The need for these clinics is widely recognized. And there isn't one progressive organization today that overlooks the importance of human relations. Why? Once again, management is primarily the function of getting the best from people with the least effort and friction. Here are six rules of human relations that will help smooth your switch to management. Start using them today—they'll save you much misery and wasted effort in any job and in your private life.

- Express and show interest in people and their problems.
- Be as impartial as you can in all dealings with people.
- Treat everyone as an individual.
- Show appreciation whenever it is deserved.
- Be firm and fair in dealing with others.
- Look for what others *can* do, not for what *you* want.

In dealing with people at work, keep their needs foremost in your mind. What do men and women want at work? Studies show that almost all of us want recognition, fair pay, security, good working conditions, agreeable supervisors, a chance to advance, important duties, and treatment as individuals. Keep these needs in view when dealing with other people. They are often the key to vague complaints, poor work, absenteeism, and other problems. As long as you are in management, your dealings with people will never end. Cultivate good human relations, and you'll solve more problems faster. See Chapter 16 for additional hints on human relations.

Learn how to delegate. To the new manager the biggest problems of his day are those related to giving orders to others. Orders must be given. But to be effective, your orders must be delivered clearly, concisely, and pleasantly. The secret of giving effective orders is to build enthusiasm in the person receiving them. You can't build enthusiasm by giving a lecture, threatening, or shouting. You build job pleasure and willingness in others in two ways: (1) by genuinely feeling this way yourself and (2) by being friendly and courteous to your associates at *all* times—not just when giving orders.

Watch the successful managers around you. Most of them are well-liked and happy people. They have a ready grin; they tackle their work with enthusiasm and gusto. Their spirit infects their associates—theirs is a happy shop. And when they give an order, it is carried out quickly and efficiently. So learn now to delegate successfully—it will be the most valuable tool in your management kit.

A new creativity. In engineering or science you cracked your mental whip to develop a new machine, solve a knotty vibration problem, or design a new process. The more creative your ideas, the greater your success as an engineer or scientist.

Much the same is true in management. But your creativity will run in different channels. Your ingenuity will help you search out the motives of other people; it will take the pulse of economic and political situations. As an engineer or scientist you may have neglected financial and business news. As a manager you will be sensitive to the many complex factors that affect your business. And the creative efforts you give to your job will include these complex elements in addition to any technical factors you must consider. The challenge of management will open a whole new field to your mind. It will stimulate and exhilarate you much like a challenging engineering or scientific problem.

Your morale will mean more. Morale on engineering and scientific projects varies considerably. In some firms the morale is high at the start of a new project. As work progresses there may be a loss of interest; sometimes there is a tedious atmosphere about the whole effort.

This may not occur in management. Here there is a steady push, a constant grappling with new and changing problems. So your morale may stay on a higher plane. And to get the best from your associates you'll try to build their morale. In doing this your own enthusiasm will increase. The atmosphere will continuously stimulate you to better effort, higher efficiency, and greater skills.

Compare the Rewards

You now have an overall view of what management work offers and what it would require of you. The rewards are real and lasting—they attract many engineers and scientists. But these rewards do not come easily—you must work to develop the skills needed to solve the constant problems management meets. Acquiring these skills will broaden your personality and widen your outlook on life.

Engineering and science, too, have their rewards. These also are real and lasting. If you are in engineering or science now, you know where you stand—your duties are familiar, and you can handle them with ease. To change to management means a completely new work atmosphere, new skills, different associates. The change is not easy; it requires much soul-searching and self-analysis. But if you're attracted to management work, give more thought to switching. These thoughts may open a new career to you. Or they may show you that engineering or science is your best career.

Test Yourself

You now have a good concept of what it's like to be a manager. Your chances of getting into management today are greater than ever. One firm, Western Supply Company, estimates that the ratio of executives to employees is currently 1 to 35; formerly this ratio was about 1 to 100. At North American Aviation, Inc., a giant of the space industry, management is technically oriented all the way to the top. At the time of the writing of this book the chairman and president of North American were both former chief engineers. Also, all division presidents, except one, had technical backgrounds.

But to get into management you must change your thinking habits. Thus, F. E. Satterthwaite, director, Statistical Engineering Institute, says: [2]

"Any engineer can handle four-variable data by intuition. Any business manager does handle five- and ten-variable data by intuition every day or he goes broke."

If you want to get into management, you've probably tried to act like a manager when you had the opportunity. But did you act like a manager at those times? Here's a three-part test developed by James M. Black [3] of Princeton, N.J., that will help determine how well qualified you are. The tests cover three qualities and skills that spell management leadership—(1) planning, (2) communicating, and (3) courage. Take these tests now and rate your skills. (Note: You can also use these tests to *rate* your performance if you are already working in management.)

TEST 1: PLANNING

Here are 10 ordinary planning situations with three alternatives for dealing with each. Check the box following each statement that best describes how you handle your job. Score yourself 10 points each time you check Statement (a); 5 points for (b); 3 points for (c). If you get 90 or above, you're either on top of planning or fooling yourself. If below, your job has you on the defensive.

1*a.* I have my assignment under finger-tip control. When my boss asks questions I'm ready with accurate and positive answers.

b. Most of the time I know the answer. But I don't like to stick my neck out, so I qualify my replies until I can check.

c. Somehow I'm always running hard just to stay even. Almost always I have to go find the answer before I can give it.

2*a.* When I attend a department conference I make sure I understand the problem involved even if it is not directly related to my work. I'm prepared to make suggestions to help in over-all planning. If I think my idea is helpful I recommend it whether or not I'm directly concerned.

b. I keep quiet unless I'm asked a direct question. I don't want to butt in.

c. I sit and, sometimes, listen. Why worry over other people's plans? I have enough troubles of my own.

3*a.* When I'm given a new assignment I plan exactly how to do it. I tell employees just what they're supposed to do and how.

b. Sometimes a new job throws me off stride. I rely on my superior to help me plan it.

c. It's hard enough carrying out the routine. When something different comes along, my boss complains about overtime, idle time, and schedule failures. But it's his job to plan, not mine.

4*a.* I know how to handle emergencies. I plan my work so emergencies seldom occur. By looking ahead and anticipating difficulties I leave myself plenty of room for maneuvering. When the unexpected happens I can deal with it.

b. I leave most of the planning to my superior, do my best to carry out his orders. If things go wrong, it's not my fault.

c. Emergencies are usual with me. But I'm not to blame. It's the boss's job to see that I have the manpower for the work.

5*a.* I give myself deadlines to accomplish projects. When given an assignment I set a time limit. I seldom fail to meet it.

b. I usually ask the boss when he wants the job finished. If he doesn't set a time, I don't. I fit the work in as I can.

c. Deadlines! I never think about them unless the work is rush. Then I let everything else go to do it.

6*a.* I plan my work flexibly so I can undertake special or emergency jobs without upsetting my regular activities. With foresight in planning I can accept an additional assignment without throwing my regular schedule out of kilter.

b. I'm usually able to get the job done. But it takes overtime and extra work if it's unexpected.

c. When extra duties come my way it's up to the boss to decide what I'm to do. I just work here.

7*a.* I know my superior bases much of his planning on the information I give him. I keep him informed.

b. I never know exactly what information the boss wants. I sometimes slip in keeping him filled in.

c. Why should I tell the boss what's going on? He should know. He gives the orders. Let him ask.

8*a.* I am familiar with the skills and abilities of my subordinates and can accurately estimate if a particular man can do a job and how long it will take. I plan to have the right man in the right spot.

b. Sometimes I have to pull a man off one job and switch him to another because my original selection was poor.

c. I work with what I have. Often I find employees idle on one job, rushed on another.

9*a.* My employees are an efficiently trained and confident working team. When I give an assignment to a subordinate I am confident it will be done without much checking from me.

b. I supervise closely. I want to make sure there are no mistakes. This keeps me bogged down in detail.

c. I seldom get around to training. Too busy correcting errors. I'm a "do-it-myself" manager.

10*a.* I constantly ask myself, "is the present method of doing a job the right one? Would better planning raise productivity?"

b. I realize it is my responsibility to plan, but usually I stick to old methods because I don't have time to experiment.

c. Why bother planning a new approach? The old way is good enough. Besides, I know it.

TEST 2: COMMUNICATIONS

Here are eight statements about attitudes in communicating upward. Take each statement in turn, apply it to yourself, check the square that tells whether it's true or false. Count up your "false" checks.

If you have eight, you're perfect. With seven marked "false," you'll do just so-so as a manager. Anything less than seven "falses" and you'll soon find yourself in real trouble.

	TRUE	FALSE
1. I build a fence around my boss. I'm a one-man protective agency, guarding him against bad news. I'm a censor of upward communications. I shield him from the facts he needs for making decisions.	____	____
2. I protect myself. I head off trouble by keeping quiet about it. I don't tell my boss about my mistakes. I "play down" facts because I know their disclosure makes me look bad.	____	____
3. I'm afraid to communicate. "Never volunteer" is my slogan. I'd rather be safe than sorry. If I tell my superior about a complication, he may give me the job of straightening it out. This may mean extra work.	____	____
4. I let other people push communications up. It's not my job to keep the boss informed. Let others worry about that. If he wants information, let him ask for it.	____	____
5. I'm not responsible, so I keep quiet. Sure, I see how a job method can be improved—an operation made more efficient. But I'm not directly concerned, and so I won't tell my superior. It's up to him to put recommendations before the right people.	____	____
6. I'm my own press agent. Upward communications is a chance for me to "blow my own horn." If I give myself a promotional boost every time I make a report, maybe the boss will think I'm as good as I think I am. Why bother about facts?	____	____
7. I always speak right up. Doesn't matter whether I know the facts. If a job is almost finished, I might as well tell him it is finished. He won't act for a while anyhow—and by the time he does act, the job will be finished.	____	____
8. I always slant the news. Usually I slant it to the positive—emphasize the good things, gloss over the bad.	____	____

TEST 3: COURAGE

If you check "yes" to Nos. 1, 2, 4, 6, 7, 12, 15 below, you have the temperament of a leader—independent, ambitious, analytical, sure of yourself. If you answer "yes" to 1, 3, 5, 8, 9, 10, 11, 13, 14, you may be a topnotch administrator or a valuable subordinate. But you don't want "make or break" authority; it's not worth the ulcer that goes with it.

	YES	NO
1. I can analyze a complicated problem and explain it in an orderly, logical way.	____	____
2. I like a difficult assignment, enjoy the "big" challenge, and am confident I can meet it.	____	____
3. I'm all right on routine decisions. But when risk is involved, I ask my superior to review my plans first.	____	____
4. I go for the hard jobs nobody else wants. If I do them well, it adds to my reputation.	____	____
5. I work best when I have my boss's approval. If he criticizes me, I'm upset, worried, and efficiency falls off.	____	____
6. I'm not afraid of responsibility, and diplomatically try to widen mine by acting on my initiative whenever possible.	____	____
7. When there are several different ways of doing a job, I have no hesitation in picking one. Usually it is right.	____	____
8. I like living by a schedule. I want my boss to lay out my responsibilities, and I stay strictly within them.	____	____
9. I'm a bug on details. Once given a job, I carry it through to the letter. I never do more or less than I'm asked.	____	____
10. I like to complete a job once I get it. Two or three simultaneous assignments confuse me. I'm no juggler.	____	____
11. I dislike delegating, am confident only when I'm doing the job myself.	____	____
12. I hate details. Give me the big problems. I'll figure out how to solve them and delegate the routine.	____	____

	YES	NO
13. Argue with my superior? Not I! He's responsible. Besides, he may get mad if I oppose him.	____	____
14. My subordinates respect me, but they know I'm the "middle man," just passing along the orders.	____	____
15. I'm willing to take the rap when things go wrong. I want credit for success, so I accept responsibility for failure.	____	____

Did your scores encourage you to give serious thought to moving into management? Or if you are in management, did the tests show that your performance is reasonably good? If the answer to either question is yes, read on.

What You May Be

"Engineers must change their attitudes if they are to become good managers," says a survey [4] conducted by Opinion Research Corporation for the Professional Engineers Conference Board for Industry in cooperation with the National Society of Professional Engineers. Deutsch & Shea, Inc., consultants in the technical manpower field, found in another survey that

> the engineer has paid for his devotion to mechanical and impersonal matters at the expense of his development as a social being. He applies far less intelligence to human relations than he does to purely technical matters, shows little interest in the social sciences, public affairs. . . . These tendencies apparently date back to the engineer's college days, when he showed a marked distaste for English and "cultural" subjects. Engineers in applications, product design, and operations are particularly cautious and conformist in personal and social relations.

What You Are—Six Assets

At an annual meeting of the American Institute of Chemical Engineers, R. L. Demmerle, assistant director of administration at General Aniline & Film Corporation, introduced a four-paper human relations session at which it was pointed out that the technically trained man:

1. Is in an enviable position to inspire trust. He is creative—he hates waste.

2. Is continually trying to push back the boundaries of knowledge. The public has great respect for such a man.
3. Is honest—almost to the point of brutality. People learn they can depend on what he says.
4. Has prestige that accompanies those who work with abstract things.
5. Is dogged in his thinking and is thus impatient when confronted with people whose thinking processes are not as well ordered and disciplined as his. (This may stand in his way in dealings with people.)
6. Is accustomed to making decisions based on fact. He lacks sympathy with the emotionally based inconsistencies of human behavior. All too often his feelings about these matters are not in the least hidden. (This, too, may stand in his way in dealings with people.) [5]

P. M. Zall, while director of the Communication Research Center at Los Angeles State College, pointed out an interesting difference between engineers and managers. Said Zall, "engineers can be distinguished from managers according to the card games they play during lunch hour. Engineers play bridge while managers prefer poker—emphasizing the cautious approach of the former and the dangerous living of the latter." [6]

21 Personal Qualifications of Managers

Many studies, tests, and surveys have been made of the traits of successful managers. Most of these studies cite many similar traits. Here are 21 traits most commonly listed: (1) sustained drive; (2) willingness to cooperate with other people; (3) original thought habits; (4) strong desires for financial security and prestige; (5) good ability to express thoughts verbally and in writing; (6) strong analytical ability; (7) willingness to take and carry out orders; (8) respect for superiors; (9) confidence in himself; (10) ability and willingness to state what he thinks; (11) persistence in seeking solutions to problems; (12) broad interests in people, situations, and nonbusiness activities; (13) ability to organize his own and others' time, effort, and energies; (14) realistic outlook; (15) decisiveness in business and personal situations; (16) knowledge of what he wants, where he's going, and how he'll get there; (17) ability to give full and immediate attention to a problem without fumbling; (18) maturity—able to act like a man at all times; (19) the drive to steadily and actively seek to accomplish more, generate new ideas, and complete tasks ahead of time; (20) the desire to emulate his superiors so he can achieve more than they have; and (21) the ability to stimulate his creative talents to raise the level of his accomplishments.

Stop for a moment and think of the successful managers you know. Compare their traits with the 21 listed above. If you spend a few minutes analyzing a successful manager, you will find he possesses many of these traits. Don't, however, expect one man to have all the traits listed. Human beings are uneven—a man who is indecisive today may, within a year, develop the trait of strong decisiveness. Also, the traits of some men lie hidden until the man is in a position where he can exert some authority.

What You Must Learn

Four top-notch managers were asked,[7] "What should an engineer learn and unlearn to become a good manager?" Their answers, in part, were:

> . . . He must think in terms of the direction of the efforts of other people rather than the completion of engineering projects; he must realize that his progress as a manager is dependent upon the number of people whom he can successfully motivate rather than the number of projects which he personally can complete. . . .
>
> There are some things which an engineer must forget when he becomes a manager—the most important of which is the exactness and accurateness with which he has appraised the results of his work. He must now realize that he is dealing with human beings. Human qualities, potentials, and relationships are anything but exact and are constantly changing. Very broad and flexible allowances must be made. On the other hand, much more is to be expected of human beings than of material things, and the life of the manager is full of many more surprises than that of the engineer.—LAWRENCE A. APPLEY, *president, American Management Association.*

> . . . All business decisions are based upon the interpretation of quantitative data, intuitive appraisals of unquantified information, consideration of the limits imposed by law and by the action of human beings in cooperative, productive effort. . . .

I would argue that the engineer must acquire a knowledge of the methods of production, distribution, and finance. He should understand as much as possible about himself, the nature of other human beings, and the environment—economic and political—in which he lives . . . —ELI SHAPIRO, *professor of finance, School of Industrial Management, Massachusetts Institute of Technology.*

There are three broad areas of knowledge and skill which an effective engineering manager must possess: an understanding of people, a knowledge of the operational functions of his company, and skill in administrating his engineering project.

A manager must accomplish his project objectives through other individuals. It is therefore imperative that he understand basic human psychology and methods of stimulating desired responses in other individuals. . . .

The engineering manager must have a clear understanding of the functions of the pertinent areas of the company, such as manufacturing, sales and service, and of his tremendous influence on their future operations. . . .

He must have skill in planning both the technical and non-technical phases of his project, as well as organizing an effective work force to achieve the technical objectives. . . .—R. E. PATTISON, *director of engineering training, International Business Machines Corporation.*

Assuming that an engineer wants to become a manager, one of the first things he must learn is how to delegate. Surprisingly, this is not always easy to do. People must be allowed freedom in doing a job and even to make mistakes. Seeing people struggle with something the "manager" is sure he could do better himself is agonizing but must often be endured.

As a manager, the engineer must develop an insight into the managerial and economic objectives of the company as a whole, acquire an understanding of the importance of programs other than his own, and maintain active cooperation with the other functions of the business.

I think we can all remember instances when engineers acted as if they were on the side lines watching the rest of the team play the game. They sometimes discuss the policies and decisions that affect their business future in a very detached manner—very critically, but detached. As long as an engineer has this attitude and does not connect his actions and contributions with those of the other players in the game, maximum contribution cannot be expected of him; and for just as long, he cannot become a successful manager.—C. F. SAVAGE, *consultant—Engineering Professional Relations, General Electric Company.*

There you have four authoritative opinions on what you should learn. Summarizing, you should learn human relations, production techniques, distribution, finance, administration, sales, service, planning, and economics.

How can you learn these subjects? This book will help you with some. For the others buy yourself at least one good book on each topic. Study these books. You'll benefit more than you ever thought possible. Never forget that the money you invest in professional books will be returned many, even hundreds—of times by the knowledge you acquire. So form the habit today of reading at least one good new book each month. Your knowledge will expand, your skills will increase, and you will derive far more joy from life. For as Thoreau wrote, "How many a man has dated a new era in his life from the reading of a book." A number of useful books covering the above topics are listed at the end of this chapter.

When choosing subjects to learn, never overlook the fact that *management is a profession—highly respected by, and satisfying to, men in it.* The importance of knowledge and dedication in a profession was strikingly pointed out in one of the most famous engineering papers of all time—"The Second Mile," by Dr. W. E. Wickenden. In his paper Dr. Wickenden wrote:

> Every calling has its mile of compulsions, its daily round of tasks and duties, its standards of honest craftsmanship, its code of man-to-man relations, which one must cover if he is to survive. Beyond that lies the mile of voluntary effort, where men strive for excellence, give unrequited service to the common good, and seek to invest their work with a wide and enduring significance. It is only in this second mile that a calling may attain to the dignity and the distinction of a profession.

Management, engineering, and science all have their two-mile road. To travel this road you must keep learning. Begin now to learn what you must know to be a proficient professional manager. For if you stop learning, you stagnate—regardless of your profession. And as an engineer or scientist don't overlook the "fourth dimension" pointed out by Morton A. Serrell, contracting engineer,

Grinnell Company, Inc. He defined this as: "The fourth dimension of the training of engineers is the training of the man to get along with his fellow man. Ask any experienced engineer. He can tell you of numbers of qualified engineers who have sought advancement in vain, been stepped over, or even fired because of lack of ability to deal with his fellow men without friction." [8]

Ten Hints on Building Your Career

Paul M. Stokes, well-known management consultant, studied the traits that rookie managers should develop to "grow up" to a full professional status.[9] Here they are:

1. Broaden your support. Stand by your subordinates. Spread your loyalty through the whole organization—up and down the line.
2. Create—don't criticize. Never squelch a new idea. Welcome the impossible—and try to make it work.
3. Shape your career pattern. Don't skitter around from job to job. Pick a pattern for your progress—and stick to it.
4. Train your subordinates. Show your self-confidence about your new position by giving others a chance to show their mettle. You'll find every working day filled with opportunities to test and train your subordinates on actual job assignments.
5. Develop your poise. Don't let the crises throw you. Every management job is full of them. Take them as they come—and keep calm.
6. Keep on learning. Don't act like a hot-shot know-it-all. Face it: there's always more for you to learn.
7. Be predictable. People want to know where they stand with the boss—today and tomorrow. Be consistent. Don't upset people by following your whims of the moment. Act so your people can be confident of your stability in making decisions.
8. Accept responsibility. Don't practice buck-passing. Reach out for more responsibility.
9. Be professional. Keep learning how to be an effective and efficient manager. Join a professional management society, attend management seminars and conferences—even if you have to pay your own expenses. Taking part in these activities exposes you to new ideas, new influences, and personal relationships that boost your management potential.
10. See causes, not symptoms. Don't panic when little causes boil

over. Diagnose them for the basic causes. Then move in and solve the problem.

REFERENCES

[1] Roger K. Crane, "What It Is to Be a Boss," *Product Engineering,* Dec. 25, 1961.

[2] F. E. Satterthwaite, private communication, Dec. 28, 1961.

[3] James M. Black, "Test Your M.M.Q.," *Factory Management and Maintenance,* April, 1960.

[4] "Career Satisfactions of Professional Engineers in Industry," Opinion Research Corp., Princeton, N.J., 1957.

[5] Gerald L. Farrar, "How to Get Ahead," *The Oil and Gas Journal,* Dec. 22, 1952.

[6] P. M. Zall, "What Every Engineer Should Know," *Product Engineering,* Dec. 5, 1960.

[7] "Points of View," *Product Engineering,* Mar. 10, 1958.

[8] Morton A. Serrell, "The Fourth Dimension," *Mechanical Engineering,* November, 1961.

[9] Paul M. Stokes, "How Savvy Managers Get That Way," *Factory Management and Maintenance,* August, 1955.

HELPFUL READING

Albers, H. H., *Organized Executive Action: Decision-making, Communication, and Leadership,* John Wiley & Sons, Inc., New York, 1961.

Buffa, E. S., *Modern Production Management,* John Wiley & Sons, Inc., New York, 1961.

Cronstedt, Val, *Engineering Management and Administration,* McGraw-Hill Book Company, Inc., New York, 1961.

Davis, Keith, *Human Relations at Work,* 2d ed., McGraw-Hill Book Company, Inc., New York, 1962.

Donaldson, E. F., *Corporate Finance,* The Ronald Press Company, New York, 1957.

Dubin, Robert, *Human Relations in Administration: The Sociology of Organization,* 2d ed., Prentice-Hall, Inc., Englewood Cliffs, N.J., 1961.

Grimshaw, Austin, and J. W. Hennessey, *Organizational Behavior,* McGraw-Hill Book Company, Inc., New York, 1960.

Heimer, R. C., *Management for Engineers,* McGraw-Hill Book Company, Inc., New York, 1958.

Ivey, P. W., and Walter Howath, *Successful Salesmanship,* 4th ed., Prentice-Hall, Inc., Englewood Cliffs, N.J., 1961.

Kierstead, B. S., *Capital, Interest and Profits,* John Wiley & Sons, Inc., New York, 1959.

Koepke, C. A., *Plant Production Control,* John Wiley & Sons, Inc., New York, 1961.

Lazo, Hector, and Arnold Corbin, *Management in Marketing,* McGraw-Hill Book Company, Inc., New York, 1961.

McGregor, Douglas, *The Human Side of Enterprise,* McGraw-Hill Book Company, Inc., New York, 1960.

MacNiece, E. H., *Production Forecasting, Planning, and Control,* John Wiley & Sons, Inc., New York, 1961.

Mauser, F. F., *Modern Marketing Management,* McGraw-Hill Book Company, Inc., New York, 1961.

Mayer, R. R., *Production Management,* McGraw-Hill Book Company, Inc., New York, 1962.

Newman, W. H., and C. E. Summer, *The Process of Management: Concepts, Behavior, and Practice,* Prentice-Hall, Inc., Englewood Cliffs, N.J., 1961.

Parker, Willard E., and R. W. Kleemeier, *Human Relations in Supervision,* McGraw-Hill Book Company, Inc., New York, 1951.

Reinfeld, N. V., *Production Control,* Prentice-Hall, Inc., Englewood Cliffs, N.J., 1959.

Shubin, J. A., *Managerial and Industrial Economics,* The Ronald Press Company, New York, 1961.

Tannenbaum, Robert, I. R. Weschler, and Fred Massarik, *Leadership and Organization,* McGraw-Hill Book Company, Inc., New York, 1961.

3
Use Your Time Better

> In time take time while time doth
> last, for time
> Is no time when time is past
>
> ANONYMOUS

As an engineer or scientist you know the importance of time in a variety of technical and scientific specialties. Thus, time is of utmost importance in launching a space probe into orbit, in the operation of certain devices in electronic circuits, in determining the rate of motion of moving bodies, and in numerous other applications. Hundreds and hundreds of your technical equations contain the lower case t or Δt, denoting that time is one of the variables in the relation.

Just as time is an important variable in technology, so too is it a major factor in your life. Proper use of time by the engineer or scientist can mean the difference between a major achievement in life and a mediocre achievement. You *can* learn to use your time better, no matter what your speciality or interests may be. In this chapter you'll discover many valuable techniques you can apply in all phases of your professional life.

Be Time-conscious

You can't begin to save time until you become more time-conscious. In becoming time-conscious you don't have to begin to run from one daily activity to another. Instead you develop a judgment for the time value of each task you must perform. Thus routine tasks not requiring your full energies can be done quickly. Creative tasks requiring your full attention and skill deserve as much time as you need to develop the desired end product.

In becoming time conscious you will realize (1) that you need some type of system to help you use your time better, (2) that you must do your work according to some plan—instead of in haphazard spurts, and (3) that the efficient use of time depends on the wise choice of the tasks you perform—thus the smart engineer or scientist doesn't try to do his secretary's work—he concentrates on his work while she does hers.

"Time is money," is the famous statement of Bulwer-Lytton. You will see later in this book that this remark is especially applicable to every engineer and scientist. For as a member of a learned profession you have many chances to consult, write, teach, or lecture—for a fee. Better organization of your time will allow you to engage in one or more of these activities in your spare time. By accomplishing more in your regular position because your time is organized better, you will be freer to pursue your outside interests no matter what they may be. Proper use of time offers so many advantages to every engineer and scientist that none can overlook the benefits to be gained. Let's see how you can begin today to get more from every hour.

Analyze Your Time Expenditures

Since "time is money,"[1] you can (1) control its expenditure and (2) budget its use. These two techniques are the keys to more effective use of your time. Saving minutes is like saving dollars—in a relatively short time the minutes build to hours just as dollars build to hundreds and thousands. This is why the wise and worldly Lord Chesterfield said, "... take care of the minutes, for the hours will take care of themselves."

To find your time expenditures, analyze your daily activities. Sketch charts like Figure 3.1. Enter in the space for each day[2] your current time usage. "But," you say, "my job is so hectic I never do the same thing on the same day in any week." Fine. Set up four or more charts, one for each week of the month. With this many charts you will certainly be able to record your activities.

Classify your tasks thus: dictation, meetings, reading, planning, manual (as in a laboratory), writing, telephoning, personal. If there are other tasks unique to your work, include them.

Block out each day in rough form to show how you currently spend your time. Be as accurate as possible. If you have trouble getting started, fill in the large blocks of time first. Thus, if you regularly attend meetings on one or more days a week, fill in these blocks first. From here move to your other regular tasks—like daily dictation and reading your mail and other incoming correspondence.

Once you have a typical week blocked out, sit back and study your time expenditures. Look at the big blocks first because it is here

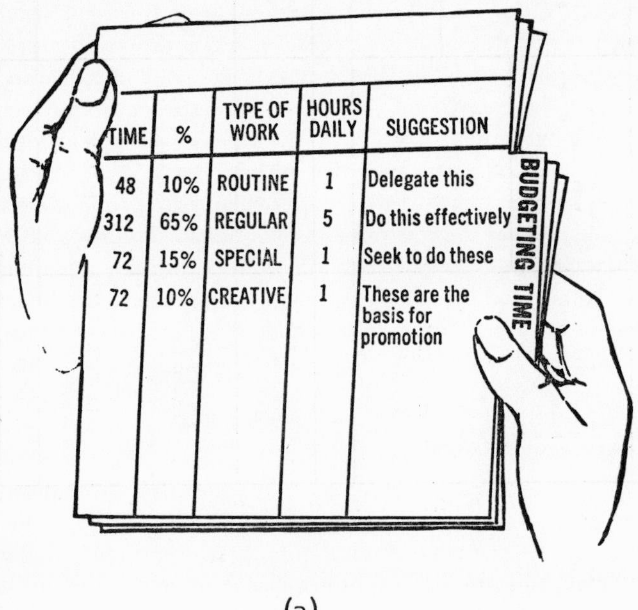

TIME	%	TYPE OF WORK	HOURS DAILY	SUGGESTION
48	10%	ROUTINE	1	Delegate this
312	65%	REGULAR	5	Do this effectively
72	15%	SPECIAL	1	Seek to do these
72	10%	CREATIVE	1	These are the basis for promotion

(a)

FIG. 3.1a. Begin your time budget by determining what types of work you do and what portion of your time they require.

	MONDAY	TUESDAY	WEDNESDAY	THURSDAY	FRIDAY
8	ROUTINE	ROUTINE	ROUTINE	ROUTINE	ROUTINE
9	Inspection and supervision of operations	Individual work with staff REGULAR	Inspection and supervision of operations	Individual work with staff REGULAR	SPECIAL WORK
10		Inspection and supervision of operations	REGULAR	Control studies and reports	Inspections and supervision of operations REGULAR
11	REGULAR	REGULAR	Division staff meeting REGULAR	REGULAR	Our staff meeting REGULAR
12	L	U	N	C	H
1	Interviews and contacts REGULAR	Interviews and contacts REGULAR	Interviews and contacts REGULAR	Interviews and contacts REGULAR	CREATIVE WORK
2	Planning and organizing	Inspection and supervision of operations	SPECIAL WORK	Inspection and supervision of operations	
3	REGULAR	REGULAR		REGULAR	
4	ROUTINE	ROUTINE	ROUTINE	ROUTINE	ROUTINE
5					

(b)

FIG. 3.1*b*. Prepare your time budget using a form like this. Supplement the time budget with daily lists of things to do. (*W. E. Dewey and Factory Management and Maintenance*)

where the greatest time savings may be possible. Do you spend a number of hours in meetings? If so, think of what you accomplish in these meetings. Does each meeting demand all your attention? If not, then perhaps you can bring other work to the meeting. While the meeting is in progress, you might read routine reports, jot the answers to letters and memos at the bottom or back of the item, fill in standard forms, etc. By judicious use of time at meetings you can often catch up on other work without missing the important items discussed.

Are there time wasters who stop by your office to chat? Do complainers take advantage of your sympathetic ear? Your time analysis will show if you're being victimized by these kinds of freeloaders. Resolve today to slash these wasteful moments to the minimum. When you see a complainer or time waster approaching, allow him or her a few moments to speak. Once you've heard the main items of his story, tell him you have a meeting or some other appointment. Gently bow him out with a promise to see him later, if necessary. The key to handling time wasters is courtesy with firmness. After a few weeks you'll find that most time wasters will respect your privacy and will come to you only when they have something of genuine importance. Then, of course, you will want to hear them out.

Streamline Your Time Budget

With your current time expenditures plotted, you can now begin to plan how to get more done every day of the year. Let's see how.

There are six tasks which probably take more of your time than any others. These are (1) reading, (2) writing, (3) dictating, (4) decision making, (5) talking, and (6) planning. You can learn to accomplish each of these tasks more rapidly if you decide to get more done each day. By consciously forcing yourself you can, like the athlete, build up speed over a period of time. Deciding today that you will increase your speed a little each day during the next three or six months can double or triple your present output, depending on your current efficiency. As an engineer or scientist you may say, "My work is too creative to be forced." This may be true, but not *all* your work is creative. By speeding your routine work you obtain more time for creative work. With increased creative hours your output can soar. So don't delay; begin today to streamline your time expenditures. As Alex Lewyt said, "Delay is responsible for more failures, more lack of success than any other circumstance."

Build reading speed. You can build your reading speed by regular practice. Should your reading speed be less than about 300 words per minute, consider taking one of the speed-reading courses available in almost every large city. If you prefer to build your speed by yourself, try these four frequently recommended hints:

(1) Use the reading guideposts provided by the author and editor—the title, subheads, illustrations, captions, tables, and other displayed material. These items give you a quick survey of the important ideas in the piece of writing. (2) Read the introductory paragraphs carefully. Pay full attention to what the author says because it is in the introduction that he makes his strongest bid to convince you that his ideas are worthwhile. Note that the usual introduction runs three or four paragraphs. (3) Next, switch to reading just the first and last sentence of each paragraph. The first sentence of the paragraph will usually state the main idea the author is discussing. In the last sentence of the paragraph the author will either summarize his main thought or will give additional evidence to back this thought. (4) Continue reading first and last sentences of paragraphs until you reach the end of the piece. Read the last paragraph completely, just as you read the introductory paragraphs. Here the author will conclude his piece. Often, he will review his earlier statements, showing why you should agree with his reasoning.

As you use this scanning technique you'll develop skill in knowing what *not* to read. For a major factor in raising your reading speed is in not wasting time on useless or uninteresting material. Practice the art of consciously urging yourself to read faster. With regular practice, or a course in rapid reading, you should be able to raise your speed from a slow 300 words per minute to a time-saving 900 words per minute.

Write faster. To write faster you must: (1) Outline before you write, even if your outline consists of only a few words jotted on a piece of paper. (2) Get over the habit of preparing several drafts of every lengthy piece. Begin today to make your first draft your final draft. (3) Write quickly, concentrating on your facts, using simple and direct words. Don't pause to try to think of words that will impress your readers. For as Norman Shidle, editor, *SAE Journal*, says, "A reader gets ideas only from words he understands. Each word in everyday writing should be understandable to every reader." Choose the words that are near at hand; most of these will be understandable to your readers.

Practice writing quickly, accurately, precisely. Cut wordiness to the bone; you'll not only write faster, your writing will be easier to read. "Know what you are going to say before you start worrying about how to say it," is another Norman Shidle recommendation. Lastly, if one way of writing—say pencil and paper—gives you more nearly a final draft on the first draft, stick to this method even though it seems old-fashioned compared to the typewriter or dictation. Better to be old-fashioned and save time than to be modern and waste time. See Chapter 9 for additional hints on writing more effectively.

Speed your dictation. To begin with, you can save time by dictating your correspondence and other written material. See Chapter 4 for hints on speeding your dictation without interfering with its accuracy. Begin today to apply those hints. Good dictating habits can save you as much as an hour per day.

Make decisions faster. Engineers and scientists probably do more dawdling before making a decision than any other group of professionals. Why? Because their training has schooled them to consider every possible alternative before reaching a decision. So you find most engineers and scientists studying every phase of a problem before making a decision. This technique is good—if you start early enough to prevent your study from delaying the actions of other people.

T. Ross Moore, president and general manager, Anglo-Newfoundland Development Company, Limited, writing in the *Top Management* Handbook [3] said, in part:

> I maintain that most intuitive things, if we understand their processes, can be learned. And decision making is largely intuitive.
>
> I have personally found one further very important thing about the decision-making process which indicates that it can be learned. The more decisions I make, the better is their quality.
>
> I have also observed that there are those who seem unable to make decisions. I have at times managed men at various levels who had difficulty in reaching decisions of any kind. Some, I find,

are merely building "a fence around their jobs." Others are actually afraid to take a chance. Still others simply do not want the responsibility of any form of decision making.

When I run into these situations, I try to peg the man immediately at the level to which he has risen. There should be no further promotions for him. A man unable to make decisions merely adds a form of paralysis to any organization.

As to what marks one man as a decision maker and another as a follower, no one really knows. It is, perhaps, buried in attitudes or basic temperament. . . .

Practice making decisions faster by obtaining needed data quickly, giving these data your immediate and full attention. Train yourself to decide on a course of action as soon as you've evaluated the facts. Then go on to the next problem without worrying about the one you've just solved. Remember Knute Rockne's famous remark, "Give me a slow quarterback with fast decisions and we'll beat most teams around." See Chapter 18 for a comprehensive discussion of other decision-making techniques that will help you save time.

Don't waste time talking. The less people think, the more they talk. You learn more, generally, while you listen or think than you do when talking. Speed your talking duties by eliminating gossip and backbiting.

To economize on your talking time, (1) start with the main idea and stick to it, (2) try to keep side issues out of the main discussion, (3) present your conclusions accurately and concisely, (4) stop the discussion as soon as the desired end point is reached, and (5) turn immediately to your next task. Remember: Economizing on talking time will help you conserve energy for other important duties.

Save time when planning. Use graphical aids like sketches, charts, photographs, and drawings to help you visualize conditions you can't see. Have your assistants collect and present data for your evaluation. Start every day with a plan—no matter how unimportant your tasks may seem. If you can't think of anything to plan, then plan on ways in which you will save time. Remember: Nothing is so empty as a day without a plan. By regular planning and execution of your plans you will become more adroit in accomplishing what you set out to do in the time available to you. See Chapter 8 for more hints on effective planning.

Begin Today

Resolve you will save time—and you will. Develop a time budget to help to evaluate your time expenditures. Don't flutter aimlessly from one task to another without completing any. Instead, plan your day and what you will accomplish—and in what order. Work at your proper professional level—don't waste time doing subprofessional work others should rightfully be doing. Don't let previous job times hold you back. It took more than a year to build an ocean-going freighter until Henry Kaiser came along. He cut the building time to less than a month—because he wouldn't let the previous normal building time hold him back. Practice making time estimates for various tasks. Then time yourself while you perform the task. You will soon develop skill in estimating how long a given task will take.

Evaluate your input time in terms of your output. Don't spend an hour on a job when ten minutes would give adequate results. Spot the circumstances in which you waste time—meetings, low-level work, unproductive people. Do something *now* about reducing the wastage. Look ahead; know what demands will be made on your time. Plan to have time available to meet these demands—but don't be too generous with your valuable time. Lastly, never kill time. Time is too precious to waste. Do something worthwhile with your time and your professional achievements will grow far beyond your fondest hopes.

REFERENCES

[1] Auren Uris, "How to Have Time for Everything," *Dun's Review and Modern Industry,* August, 1957.

[2] W. E. Dewey, "You Can Use Your Time Better," *Factory Management and Maintenance,* February, 1953.

[3] H. B. Maynard (ed.), *Top Management Handbook,* McGraw-Hill Book Company, Inc., New York, 1960.

HELPFUL READING

Cooper, J. D., *How to Get More Done in Less Time,* Doubleday & Company, Inc., Garden City, N.Y., 1962.

Josephs, Ray, *How to Gain an Extra Hour Every Day,* E. P. Dutton & Co., Inc., New York, 1955.

Laird, D. A., *The Technique of Getting Things Done,* McGraw-Hill Book Company, Inc., New York, 1947.

Leedy, P. D., *Reading Improvement for Adults,* McGraw-Hill Book Company, Inc., New York, 1956.

Uris, Auren, *The Efficient Executive,* McGraw-Hill Book Company, Inc., New York, 1957.

4

Organize Your Workday for Efficiency

> Order: Let all things have their places. Let each part of your business have its time.
>
> BEN FRANKLIN

No matter what you choose as your lifetime specialty—engineering, science, or management—you will accomplish more if you organize your workday for efficiency. You never lose when you tailor your efforts to produce a greater output. Ted R. Arthur, while an executive of Byron Jackson Pumps, Inc., said, "... the more we develop ourselves to secure the maximum of results with a minimum of effort, the closer we are to the ultimate of energy conservation, through the process of economy, precision, and effectiveness of action."

How do you organize for higher efficiency? It's easy. All you need do is analyze your work load. For most engineers and scientists the work load can be divided into five categories—*paper work, talking, writing, reading,* and *miscellaneous.* Analyze your own daily activities. You will find that these categories are accurate, unless you spend a major part of your time in experimental work in a laboratory or test room. If this is so, then you'll have to add a sixth category—*manual tasks.* Let's take a look at each category to learn how you can streamline your efforts so you accomplish more with less effort.

Paper Work Can Be Streamlined

Don't groan at the thought of the papers you must juggle to earn a living. All of us in business today are beset by the same problem. Yet some men seem to be able to handle paper work more readily than others. Why? Because these men are better organized. They know how to handle papers quickly and efficiently. With a little practice you can do the same. Your engineering or scientific training will help you solve the paper-work problem with a minimum effort. Here are twelve keys to efficiency:

1. Start with the top of your desk. Become a *clean-desk man*—that is, keep the desk top cleared for action. Pile your papers in a neat stack every morning. Arrange the stack so the most important papers are on top, the least important on the bottom.

2. Begin your day by taking action on each piece of paper in the stack. Start at the top and work toward the bottom. "But," you say, "I can't work on three of these items until I have more data." Fine—have those three items filed until the needed data arrive. By putting these papers in the file until you can take action on them, you eliminate the need to think about them. Your stack of paper work grows smaller, allowing you to give more attention to the remaining items.

3. Take action immediately. If a piece of paper work requires an answer—say a letter or memo—jot a short outline of your proposed answer at the bottom or on the back of the paper. Thus, you might make this note: "Thanks—will send." In dictating your answer you'll expand these three words into a full-fledged letter. But your short outline will help you dictate more rapidly and accurately because the general content of the letter will be in your mind.

4. Use longhand notes to answer memos and other interoffice correspondence that doesn't require the formality of a typewritten answer. You can even retain carbons by using the standard snap-type memo forms having carbon paper attached to them. Analyze your interoffice and interplant correspondence, and you'll find that much of it can be done by handwritten notes. Try longhand notes today—you'll find they permit you to take fast, direct action, save your secretary's time, and reduce that pile of paper work in minutes, instead of hours. Of course, you must use judgment as to whom you

send handwritten notes to. In general, don't send a longhand memo to top management unless this form is approved by your organization. Incidentally, by using longhand memos you save your organization money. Studies show that the average cost of a typewritten memo today is about $1.72.

A longhand memo also lends a personal touch to your message, particularly when the memo goes to someone who is at a lower level in the organization. Some executives make it a regular practice to send all congratulatory memos in longhand. The recipient of the memo is not only delighted by the contents—he is also pleased that his superior took time out to write the message himself. That's why you'll sometimes see an important longhand memo framed in glass for the recipient and all his family and friends to admire.

Lastly, you might consider a general rule followed by some engineers and scientists: Do not, in general, write a longhand reply to a typed communication from top management. The longhand form may be considered unbusinesslike by people at the top of the management pyramid.

5. Arrange paper work ready for dictation in an order of decreasing importance. Keep this paper work in a separate folder, ready for your next dictating session with your secretary or dictating machine.

6. Form the habit of delivering some outgoing paper work in person. By walking to the place where the papers are to be delivered, you obtain healthful exercise, preventing you from becoming, as Ted R. Arthur says, a "swivel-chair jockey." You'll meet people and see things during your walk. This keeps you in closer touch with what's going on. And getting away from your desk renews your perspective, relaxes your mind, and provides you with a welcome break in the day's routine. You'll also have a chance to discuss the contents of the papers with the recipient. An oral face-to-face discussion will often enable you to settle a complex matter quickly and efficiently.

7. Secretary tied up? Don't fret—write your letters in longhand or type them yourself if you can type neatly and clearly. Doing your letters either way will help you keep caught up. It will also teach you the value of brevity. For when writing in longhand or typing, you will weigh the importance of each word. Your letters will be more direct and easier to read. Train your secretary or transcribing

department to work from a longhand letter that begins "Dear Mr. ——," and ends "Vty." Attach related correspondence to the longhand letter. Then the typist can pick up the name and address of the addressee. There's no need for you to copy the name, title, address, and other data in longhand.

8. Try to work to the bottom of your paper-work pile before noon. This will free your afternoon for creative work and other duties. With a clean desk top you'll work faster and more surely.

9. Keep technical and scientific magazines, books, catalogs, and similar publications out of your paper-work stack. Arrange all publications on a separate shelf, out of the paper-work line of fire. This permits you to concentrate on your paper work without having to shuffle publications which may distract your attention.

10. Use "to do" lists to prevent forgetting important items. List paper-work jobs in an order of decreasing importance. Cross off each job when finished. This gives you a feeling of accomplishment as you see the list decrease in length.

11. Learn to scan paper work while you're on the phone. Many phone calls are routine in nature. By giving half your attention to the phone and half to your paper work you can accomplish more in less time.

12. Never allow routine paper work to build into a backlog. Say you have certain daily reports to read. Set aside a few minutes each day, preferably in the morning, to read these reports. Make the reading of these reports as important as your coffee break. Then you'll be sure to stay one jump ahead of your routine paper work.

Use these hints now. They'll save you time and will help break the paper-work logjam. Alter the hints to suit your personal situation—this will enable you to be a clean-desk man at all times.

Lastly, develop respect for the usefulness of your wastepaper basket. Don't save papers for later reference unless they're important. Instead, develop a good memory for items you read and want to use. Toss the paper into the wastepaper basket after you've read it. That way you'll unclutter your desk and your life. Paper collectors seldom reach the top in any job.

Use a diary to schedule your time. Get a neat, small diary for the top of your clean desk. The Nascon "Week-at-a-Glance" is excellent because it shows you the next seven days. Avoid tear-off type calendars unless you are particularly fond of them. The tear-off type doesn't allow you to see ahead without extensive shuffling of pages. Also, every time you tear off a page and throw it away, you destroy a valuable record.

Keep your desk diary neat and clean at all times. Make entries only in ink. Once you've finished a task noted in the diary, check it off with a red pencil. The contrast between entries in blue ink and red checks tells you at a glance what was accomplished.

Be sure your diary has a permanent binding. Then you won't lose pages from it. The diary is useful in maintaining a record of your daily activities. With neat, accurate entries the diary can help you support business tax claims you may have. One caution with weekly diaries—be sure to make entries in the proper week. Some diaries are arranged so the lower left corner of the lefthand page can be clipped off to identify the week. Clip this page every Monday morning before you start work. Then you won't make notes in the wrong week.

Make Talk Pay Off

C. L. Brisley, Ph.D., of Wayne State University, studied the daily activities of technical operating managers over a period of several months. His conclusion: "Talk. Talk. Talk. That's a quick summary of the manager's life. No question, the largest chunk of his time is spent in communication." Typical top and operating managers spent over 80 per cent of their time, or about 110 hours per month, talking. This talking took place in consultations, interviews, telephone calls, dictation, meetings, etc. So the sooner you make business talk pay off, the quicker you can improve your workday efficiency. Here are ten hints for improving your business and technical talking chores:

1. Cut trivia to the minimum. Learn how to size up people with whom you talk. Spot the windy explainer, the detail prober, the storyteller, and the other earbusters who want to waste your time. Hear them out for a few moments—then shift the conversation to specifics by asking, "What did you want to see me about, Joe?" Or try, "What's today's topic, Sam?" With a little practice you can develop a number of other opening sentences that will swing the conversation from trivia to specific topics.

2. Keep the conversation on the track. Don't allow the talker to wander. If he switches to his wife's PTA activities, swing him back with something like, "Very interesting, Joe. I'd like to hear all about it at lunch someday. Now what were you saying about the new design?" If you keep the talker on the subject, be sure you also stick to the topic. Avoid stories about the past, other jobs, and other people, *unless* they are relevant to the main topic of conversation. If you *must* tell a story or use a simile, limit yourself to one. This way you play fair with your listener and keep the conversation on the track.

3. Limit telephone talk to essentials. Avoid golf, hunting, boating, and similar hobby talk on the phone. This kind of talk only ties up valuable switchboard time, wasting minutes and energy at both ends of the wire. Get yourself one of those small three-minute time glasses or one of the new mechanical timers. Though these are made to time long-distance calls, they are equally useful for local calls. Try to limit the duration of your local calls to five minutes or less. To close off lengthy, unimportant telephone calls, learn to inject a note of urgency into your voice. With a little practice you can close off a conversation quickly and politely. Of course, you can also use the dodge, "Joe, is that all? I have a long-distance call on the other wire." This dodge is even used by men who do not have a two-extension phone.

4. Talk to people on your terms—not theirs. By calling a person instead of waiting for him to call you, the conversation can be streamlined—*in your favor*. Go to another man's office if you want to speak to him personally. This puts you in control of the conversation. You can open the discussion quickly, come to the point, get the facts, and close the conversation by leaving to "see someone else," "get the work started," etc.

5. Keep tight control of talk at meetings you run. This doesn't mean you should try to stifle creative talk at meetings. What it does mean is that you should reduce trivial talk to a minimum. For the more you allow people to depart from the subject of the meeting, the less you accomplish. People at meetings have difficulty bringing their attention back to the main topic after their minds have been diverted by unrelated talk. Useless talk at meetings distracts everyone present, leading to a major waste of time. One mark of a good meeting chairman is the ability to limit loose talk.

6. Be careful to be friendly with people close to you in your organization. Your secretary, your associates, your boss—all these people are interested in *you* as a person. And you should be interested in *them* as persons. So don't make the mistake of trying to hide every aspect of your personality from them. But don't spend hours discussing personal matters. An occasional inquiry about a man's family, his hobbies, and his other activities is enough. Just be sure to remember what you hear so you can continue the discussion at some time in the future.

7. Shun gossip, slander, backbiting, and other degrading forms of talk. You seldom gain when you engage in talk harmful to another person.

8. Listen more than you talk. You have less chance to learn something while you're talking. By listening you give the other man a better chance to tell you what's on his mind. Also, the less you say, the more meaningful will be your guiding remarks that help keep the conversation on the track.

9. Have your secretary or associates screen incoming phone calls when you're busy. This will reduce interruptions, allowing you to finish your work sooner.

10. If you have a secretary, teach her to place long-distance calls for you. You'll save time and give her a feeling of deserved importance.

By controlling your talk and the talk of others, you can measurably improve your workday efficiency. So begin now to apply the ten hints given above. Once you've put these hints into action, you will be better equipped to streamline another important talking chore—dictation.

Dictate—you'll save time. Practice, preparation, precision, and persistence are the keys to effective dictation. By learning to think in terms that will permit you to dictate clear and accurate thoughts, you can save many hours every week. If you've avoided dictating technical reports, papers, articles, and similar written pieces, reconsider your decision. With a little practice you can learn to dictate your thoughts in much less time than you can write them in longhand or on a typewriter.

To develop your dictating ability, begin with short letters and memos. Study each letter or memo in advance, making a short outline of key words to guide your dictation. Don't try to dictate a letter or memo until after you know what you intend to say. When you're dictating to a secretary, have a clear idea of what you'll say in each letter or memo *before* you call her in. Use the same technique with a dictating machine. By devoting a period to thinking and a period to dictation, you will improve the results obtained. Don't spend a few minutes thinking, a few minutes dictating, and then switch back to thinking. The sudden changes in thought patterns will lead to lost motion and wasted time.

Practice dictating daily. If you don't have a secretary, or you'd prefer her to be unaware of your practice, get a dictating machine. One advantage of the machine is that it permits you to play back your dictated words. By listening critically to the playback you can detect unclear thoughts, poor enunciation, and sloppy organization of ideas. When you switch from the machine to a secretary, keep one thought in mind—your secretary is probably as nervous as you are for the first few sessions of dictation. So make her relax by getting her to laugh. You'll laugh too, and both of you will be more at ease.

Prepare—in advance of every dictating session. Where you know the man to whom you're writing, visualize that he's sitting across from you while you're dictating. Speak to him in clear, concise terms. Give him the needed information and stop. Just because you have a microphone in your hand or your secretary is sitting opposite you is no reason for becoming wordy. Keep dictated material short and to the point. Use lively words that will awaken response in your reader's mind.

Be precise at all times. In long letters where you refer to "the above items," name the items, even if it seems repetitious. Being precise in letters will help make your reports and technical papers more accurate and useful. Preciseness will keep you out of trouble and will help build your reputation as a talented and dependable engineer or scientist.

Be persistent in practicing dictation. Once you've mastered letters and memos, turn to reports, articles, papers, etc. Have the typist triple-space the first few items of this kind that you dictate. Then you'll have plenty of room to make corrections. As your skill develops, you can switch to double-spacing for the first draft. As with any other form of dictation, preparation is of prime importance for reports, articles, papers, and similar items. So don't worry if you spend more time preparing to dictate than you do dictating. This is typical of a person who gets maximum output from his dictation time. Remember: *Persistent practice pays dividends in dictation.*

Here is an excellent summary of good dictation practice, as recommended by Robert L. Shurter, professor of English at Case Institute of Technology:

> (1) Relax and dictate in a clear, natural tone of voice. (2) Enunciate clearly. (3) Spell out any names or words that sound similar to other names or words. (4) Dictate only periods, paragraphs, and unusual punctuation. Leave normal punctuation to the transcriber. (5) Ask the transcriber to read back—or if it is machine dictation, play back—any parts of the message which may not be clear.
>
> In machine dictation (1) Hold the microphone or speaking tube in the position recommended by the manufacturer. Don't wave it around; don't shift it from one position to another. (2) Speak in a normal conversational tone. Don't shout and don't drop your voice. (3) Talk steadily and at a little slower rate than normal, unless you are a very slow speaker. Don't fluctuate by rushing occasional phrases or by hesitating for long intervals. (4) Take special care to enunciate clearly, to speak distinctly, and to pronounce unusual words with unusual care. If there is any doubt, always spell such words out. Don't mumble. (5) Dictate paragraphs and uncommon punctuation: semicolons, colons, dashes, parentheses. Voice inflections, as in normal conversation, will indicate commas and other conventional punctuation. (6) Dictate figures by digits. (7) Make all corrections carefully in accordance with the directions for the type of equipment you are using. (8) Attach all necessary forms, guide sheets, cards, and all other information to the cylinder or records. (9) Start the machine before you begin dictating and stop it after you have finished.[1]

Read for Profit

As a clean-desk man you'll have your current reading matter—technical or scientific magazines, journals, catalogs, books—neatly arranged for ready reference. Turn to these regularly to learn the latest developments in your field and to give a change of pace to your day. The more technical, business, and scientific magazines you read, the wider will your knowledge become. Alert reading of current journals will give you many new ideas, stimulating your interest and creativity

Here are ten hints *Fleet Owner* magazine gives to help you get more out of any periodical you read:

> (1) Take a first, fast look at the contents page—in most magazines this page is located in the same position every month, so learn where it is. (2) Look for articles about your own specialty because they'll be of immediate help to you. (3) Be on the lookout for ideas at all times. But don't expect to lift them "as is." The trick is to *adapt before you adopt*. (4) Read about the other fellow's job . . . you'll get a wider view of your field. (5) Use all the magazine's service departments—they'll keep you up-to-date on special subjects. (6) File tear sheets or get reprints of articles that specially interest you. (Note: Many technical and scientific magazines will send you a free tear sheet of an article. Some organizations have special post cards printed to be used in requesting tear sheets.) (7) Save back issues of important magazines and journals for a year or more. These issues will help you build a permanent file of valuable information. At regular intervals, clip and discard back issues so your shelf does not become overcrowded. (8) Write the editor when you have technical problems. Your letter will help him plan future issues of particular value to you. (9) Read the advertisements. They'll keep you posted on new developments in your field. (10) Write to the manufacturers for information on equipment of interest to you. The descriptive literature you receive can be extremely useful.

Learn to scan technical and scientific publications. This will help you read more periodicals in less time. Remember: The greater the number of publications you see, the more ideas you can acquire. If you have an intelligent assistant, try the scheme Rogers, Slade and Hill, New York consultants, report one successful manager uses: He flips through each magazine or report as it comes in. From headlines, subheads, and lead sentences he can tell immediately those that are important to him. He marks them for his assistant. The assistant then clips the marked pieces, underlines key thoughts, and returns the clippings to the manager. Thus marked, the substance can be quickly extracted and the clippings filed, saved for complete reading at a later date, or thrown away.

This procedure takes about a half hour weekly and accomplishes three objectives. It saves you from plowing through every page and line of a magazine or journal. It leaves your evenings at home free. And it permits you to choose what is important without delegating responsibility to a subordinate who may not have your awareness.

Learn to read some general, nontechnical business magazines also. Thus, you might try *Business Week, Fortune, Dun's Review and Modern Industry, Think, Supervision, Barron's,* etc. These magazines will help you get a wider picture of the business side of science and technology. For example, a recent issue of *Factory* has these hints for reducing the time lost in small talk: Jangling telephones and casual visitors, whose only purpose is to maintain contacts, are among the greatest killers of an executive's time. So says Joseph M. Trickett, University of Santa Clara (California) business administration professor.

If your runaway minutes are turning into nonproductive hours, Dr. Trickett offers these five hints to help you to better control your time.

> 1. Try to do only one thing at a time. Don't spend the working day doing a bit of this and a bit of that.
>
> 2. Complete each action. When going through a stack of mail, you should not lay a letter down until you decide on some course of action.
>
> 3. Your office layout should fit your needs.
>
> 4. Take your secretary into your confidence. She can filter phone calls and sort your mail.
>
> 5. Set a specific time for subordinate contacts. Let them have a designated uninterrupted period of discussion. Also schedule group sessions.[2]

Don't consider the time you spend on business magazines wasted. The articles, departments, and advertisements will widen your horizons. Regular reading of good business magazines helps you understand more clearly the place of your field in the general economy. Once this is clear in your mind, you will understand how your job contributes to better living for all.

Test Your Deskmanship

Once you've applied the hints in this chapter, check on yourself. Score yourself on the quiz in Figure 4.1. This quiz[3] was developed

HOW'S YOUR DESKMANSHIP?

MAKE THIS SIMPLE CHECK, FIND OUT YOURSELF

		PENALTY POINTS PER UNIT	NO. OF UNITS	TALLY
1. For each piece of today's mail still in your in-box at 5 p.m.		5		
2. For each piece of mail not answered	in 3 days	2		
	in 1 week	10		
	in 1 month	50		
3. For each 1-in. thickness of files or unclassified papers on your desk top not used last week		2		
4. For each unread magazine, periodical, technical paper, or brochure on your desk top		5		
5. For each letter you wrote last month beginning, "I'm sorry for the delay in answering your request"		10		
6. For each letter you received last month beginning, "We have received no answer to our letter..."		50		
7. For each time you couldn't locate a piece of correspondence last month		10		
8. For each completed item in your desk files over 1 year old		½ (max 20)		
9. For each requested report you receive but don't need or use		50		
10. For each "miscellaneous" file	over 1/2 inch thick	5		
	over 1 inch thick	25		
	over 2 inches thick	50		
			TOTAL	
			DEDUCT	
			ADD	
			FINAL TOTAL	

HANDICAPPING
Deduct for supervisory load
25 points if you supervise over 50 people
50 points if you supervise over 100 people
100 points if you supervise over 500 people
Add for secretarial help
25 points if you have a half-time secretary
50 points if you have a full-time secretary

SCORING	
less than 50	Unbelievable
50 to 100	Darn good
100 to 200	Time to get busy
200 to 400	Calls for drastic action
over 400	You're snowed under

FIG. 4.1. Check your deskmanship using this easy test. Score your skills and then get busy to improve, if necessary. (*Robert B. Wilson and Factory Management and Maintenance*)

by Robert B. Wilson, while assistant vice president of Wallace Clark & Company, Inc.

If your quiz score is high, don't sit back and relax. Keep trying to devise new ways to improve your working efficiency. By giving more thought to your job you will increase your creativity, raise your output, and achieve more in your chosen field.

As a final check on your new working habits make a regular practice of checking your "in" basket. If it's empty—or nearly empty—you're doing well. But don't stop there. Pull open the drawers in your desk and files. If they make you shudder, if you don't recognize what's in them, it's time to act. Apply the hints in this chapter more rigorously.

REFERENCES

[1] Robert L. Shurter, "Written Communication in Business," McGraw-Hill Book Company, Inc., New York, 1957.

[2] "Small Talk Kills the Business Day," *Factory Management and Maintenance,* November, 1962.

[3] Robert B. Wilson, "How You—Yes, Even You—Can Be Boss of Your Desk," *Factory Management and Maintenance,* September, 1956.

HELPFUL READING

Bender, J. F., *How to Talk Well,* McGraw-Hill Book Company, Inc., New York, 1949.

Brown, Leland, *Communicating Facts and Ideas in Business,* Prentice-Hall, Inc., Englewood Cliffs, N.J., 1961.

Buckley, E. A., *How to Write Better Business Letters,* 4th ed., McGraw-Hill Book Company, Inc., New York, 1957.

Bury, Charles, *Telephone Techniques,* Charles Bury and Associates, Dallas, Tex., 1955.

DeVoe, Merrill, *How to Use the Telephone Profitably,* Prentice-Hall, Inc., Englewood Cliffs, N.J., 1955.

Dietrich, J. E., and Keith Brooks, *Practical Speaking for the Technical Man,* Prentice-Hall, Inc., Englewood Cliffs, N.J., 1958.

Ingram, K. C., *Talk That Gets Results,* McGraw-Hill Book Company, Inc., New York, 1957.

Laird, D. A., *The Technique of Getting Things Done,* McGraw-Hill Book Company, Inc., New York, 1947.

Leedy, P. D., *Read with Speed and Precision,* McGraw-Hill Book Company, Inc., New York, 1963.

Murphy, Dennis, *Better Business Communication,* McGraw-Hill Book Company, Inc., New York, 1957.

Parkhurst, C. C., *Business Communications for Better Human Relations,* Prentice-Hall, Inc., Englewood Cliffs, N.J., 1961.

Schwartz, Jack, *How to Get More Business by Telephone,* The Business Bourse, New York, 1953.

Shurter, R. L., *Effective Letters in Business,* McGraw-Hill Book Company, Inc., New York, 1954.

Surles, Lynn, and W. A. Stanbury, Jr., *The Art of Persuasive Talking,* McGraw-Hill Book Company, Inc., New York, 1960.

Uris, Auren, *Developing Your Executive Skills,* McGraw-Hill Book Company, Inc., New York, 1955.

Whiting, P. H., *How to Speak and Write with Humor,* McGraw-Hill Book Company, Inc., New York, 1959.

5

Creative Thinking Pays Off

> Let me exhort everyone to do their utmost to think outside and beyond our present circle of ideas. For every idea gained is a hundred years of slavery remitted.
>
> RICHARD JEFFERIES

As an engineer or scientist you may be able to *triple* your output of good ideas—after you've had some training in creative thinking. For example, at General Electric Company, where engineers have had courses in creative thinking available for years, the men who've taken these courses can produce *three times* as many good ideas as those who haven't. Can *you* improve your creative thinking abilities? Yes, if you try. Research and studies show that the average engineer or scientist can improve his idea output. Let's see how.

What Is Creative Thinking?

Creative thinking in engineering or science is the ability of an individual to repeatedly conceive and develop new and useful results by using his imagination, past experience, memory, or other abilities. The word creativity has been so broadly applied in recent years that it can represent any of four steps[1] in science and engineering: (1) pure and basic research of the Nobel Prize level, (2) innovation and discovery, (3) invention, and (4) problem solving. The exact level at which you work will vary during your career. For example, Figure 5.1 shows the number of patent applications made in the radio field by 112 inventors at a research laboratory during a six-year period. Note the age distribution for this particular group. The results obtained in the study of this group of inventors agree, in the main, with more extensive studies of age groups for invention in all fields.[2]

Recent investigations in the field of creative thinking show that academic, industrial, and government leaders are acquiring greater knowledge of and respect for creative thinkers. By building your creative abilities you can achieve more in your profession. And the climate for creative thinkers was never better than it is today.

Don't confuse creative thinking with other mental processes—judgment, reflection, reasoning, remembering, observation. These processes use what Alex F. Osborn (well-known writer on creativity and an originator of "brainstorming") calls the judicial side of your mind. The other, or creative side, of your mind is responsible for generating new and useful ideas. Your creative powers are the key

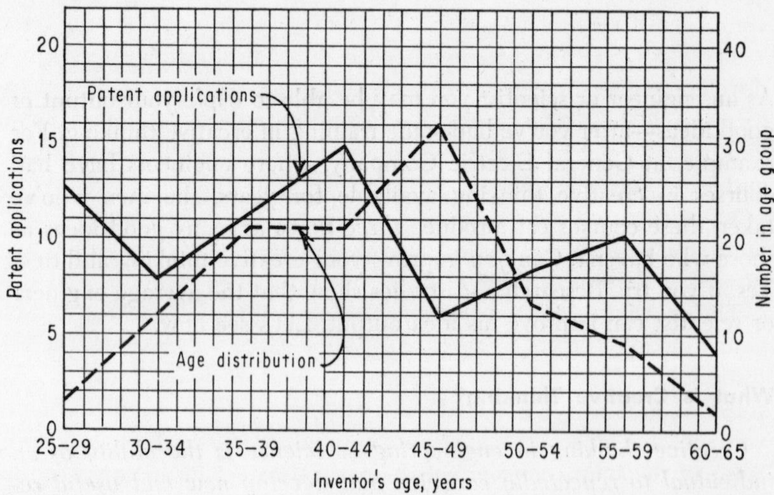

FIG. 5.1. Inventor's age versus patent applications. Note the age groups in which applications are high. (*C. D. Tuska, Inventors and Inventions, McGraw-Hill Book Company, Inc., New York, 1957*)

to bringing original ideas into being. Once you've developed some original ideas you can evaluate their worth, using the judicial side of your mind. But if you let the judicial skill, which is usually strongest, work while you're trying to be creative, your creative efforts may be wasted. That's why it is important that you train yourself in the techniques of creative thinking. Knowing these techniques will help you produce more good ideas in less time.

Steps in Developing Creativity

There are two important steps you must take if you want to improve your creative ability. Dr. J. P. Guilford, while director of the Psychology Laboratory, University of Southern California, described these steps as: (1) acquire an understanding of the process of creative thinking and the factors that affect it, and (2) use the right kind of practice to improve your creative ability.

Further, Dr. Guilford offers the following encouraging facts: Everyone can be creative to some degree in some ways. For some people, the best creative outlet is mechanical; for others it's organizing—or gardening—or painting. But, by applying the creative methods you use in your specialty to other problems, you'll increase your overall effectiveness.

Table 5.1 is a useful summary of three approaches to creative development of ideas. John Dewey, the great American philosopher, formulated the *normal thought process* for solving problems. Here you begin by determining the real problem and then proceed to fact gathering, obstacle analysis, solutions list, and plan of action.

The *structured approach*, used in General Electric's courses, uses normal thought processes and follows a predetermined patterned approach or formula. Table 5.1 shows one form of the pattern used.

The *free-wheeling* approach, pioneered by Alex Osborn, allows your thinking to work unchecked. It utilizes the sudden idea, the intuition, the chance guess. You set no limit to your thinking. Instead of limiting yourself to conventional solutions for a problem, you seek out every practical and impractical answer you can find. Some of the solutions—perhaps many of them—will be unworkable. But some may be highly practical. People who tackle problems this way say that 6 to 10 per cent of the ideas turned up are usable.

Widely used, Osborn's *free-wheeling applied-imagination* ap-

Table 5.1 THREE FAMOUS FORMULAS FOR THINKING UP IDEAS

John Dewey's NORMAL (*Normal Thought Process*)	General Electric's STRUCTURED (*Patterned Approach*)	Alex F. Osborn's FREE-WHEELING (*Applied Imagination Sequence*)
1. Determine the real problem	1. Recognize 2. Define	1. Orient—point up problem
2. Gather all facts	3. Search	2. Prepare—gather pertinent data
3. Analyze the problem—obstacles to overcome	4. Evaluate	3. Analyze—break down relevant material
4. List possible solutions for overcoming obstacles	5. Select 6. Make preliminary design 7. Test and evaluate	4. Hypothesize—pile up alternatives by way of ideas (brainstorm) 5. Incubate—let up to invite illumination 6. Synthesize—put the pieces together
5. Develop plan of action	8. Follow through	7. Verify—judge resultant ideas

SOURCE: Lester R. Bittel, "How to Make Good Ideas Come Easy," *Factory Management and Maintenance,* March, 1956.

proach is unique. In his book, *Applied Imagination,* Osborn points out:

> In actual practice we rarely follow such a 1-2-3 sequence (as outlined in Table 5.1). We may start guessing while we are still preparing. Our analysis may lead us straight to the solution. After incubation we may again go digging for facts which, at the start, we did not know we needed. And, of course, we might bring verification to bear on our hypothesis, thus cull our wild stabs and proceed only with the likeliest. . . . All along the way we must change pace. We push and then coast, and then push.[4]

This is true of any of the systematic approaches. Remember that creative thinking is not always logical thinking. To up your creative output you must forget step-by-step procedures used in the solving of design and numerical problems. Realization of the differences between creative and analytical thinking is one of your biggest steps toward improving your creativity.

Start Now to Build Your Creative Powers

Here are eight useful principles to unleash your creative powers. Begin to apply them now—in as many of your activities as you can. Don't wait to practice these principles on puzzles or games. Use them on your job where they will pay dividends.

1. ***Define your problem.*** Be specific; write the problem out as concisely as possible. Thus, your problem might be: *Cause of failure of electronic semiconductor material ZN31,* or *Low-cost device for thrust stabilization.* If you can't define your problem, you can hardly expect to be creative about it. And without creativity your hopes for a useful solution are nil.

2. ***Focus your attention on the problem.*** Put other problems and thoughts out of your mind. Take aim on your present problem and devote all your energies to its solution. Don't scatter your thoughts—concentrate them on the problem you defined in item 1.

3. ***Create first, judge second.*** Open your mind to the problem—freewheel, making a list of every solution you can find. Let's say your problem is poor performance of a new electronic semiconductor. List every possible cause of the trouble—faulty manufacturing techniques, poor materials, wrong assembly methods, etc. Try to obtain a dozen or more possible causes by the freewheeling method. Then sit down and analyze the list of causes. Choose those you judge to be the most probable. Switch to your creative approach and develop as many facts about the probable causes as you can. Then judge these facts. Continue alternating between creative freewheeling and analytical judgment until you have an acceptable solution.

4. ***Continue to develop ideas.*** Once your creative ideas begin to flow, keep them coming. One idea will suggest another. Jot each idea down as fast as it comes. Don't stop until you feel a definite sense of letdown or completion. For if you stop before this feeling sets in, it may be almost impossible to start your ideas flowing again. Respect your creative powers—coddle them while they are in full swing.

5. ***Don't give up—continue thinking.*** You will often be discouraged because ideas won't start flowing. Continue searching, in your mind, for that first idea. Once this comes, others will follow. So persevere—worthwhile ideas seldom come easily.

6. *Take an idea break.* Stop thinking when fatigue sets in. Allow your subconscious mind to take on the problem. Turn to other tasks. Then, at the most unexpected moment the solution may occur to you. Thousands of engineers and scientists have experienced this. Poincaré's conception of Fuchsian functions came to him while he was boarding a bus; Armstrong's understanding of superregeneration developed while he was using some measuring apparatus; Watt conceived of the steam condenser while strolling on the green of Glasgow.

7. *Have confidence in your creative abilities.* Believe that you can develop good ideas—then work at bringing these ideas into being. Confidence in yourself will do much to increase your creative output. Actively seek tasks that demand creative thinking. Apply your imagination to them. Be constantly alert for sudden solutions that will pop into your mind. Keep a sharp eye for ideas that will occur to you while you are relaxing—after you've turned your mind to other matters.

8. Take action on your ideas. Don't file good ideas away for later use. Make a list of all your useful ideas and go to work on them. Remember: Idle ideas benefit no one. Take action on each idea. Seeing your ideas successfully put to work will stimulate you to greater creativity.

Convince Yourself Your Ideas Are Good

Getting ideas is only half of any creative task. The other half of your job is to put your new ideas to work. To do this successfully you must first convince yourself that the new ideas are worthwhile. Strange as it may seem, some creative people meet more resistance within themselves to new ideas than they do from their associates or supervisors. Here are typical idea roadblocks you may encounter within yourself.

Yours is an "it won't work" mind. Some engineers and scientists are pessimistic about and critical of almost every new idea. They see ten or more negatives for every positive. Sometimes these men are termed the "it won't work boys." This term is applied to them because their most common remark begins, "It won't work because. . . ."

Recognize now that new ideas will often violate existing practices, may flout statistics, and may appear impractical at first glance. If you take an immediate it won't work attitude toward new and off-beat ideas you will seldom benefit from your creative efforts. So every time you feel an it won't work reason ready to roll off your tongue, restrain yourself. Think about the idea; if possible try it in actual practice. Ignore statistics for awhile, if doing so does not endanger health or property. Sometimes the idea that "won't work" is the most successful technique introduced in years.

Ridding yourself of the it won't work attitude has other advantages. Extreme pessimism and a critical attitude can warp your personality. We've all met engineers and scientists who were so critical of almost every part of life that they were depressing to be near. Men with pessimistic and critical attitudes seldom achieve much—they are too busy looking for things to criticize. By taking a more optimistic outlook on your profession you give things a chance to happen. Before you know it one of your seemingly won't work ideas *does* work. One really good idea can mean the difference between outstanding achievement and a routine career.

Your emotions block you. Emotions can hinder many creative activities. Thus, if you [3] *fear* people who laugh at inept ideas, *worry* about what others think of you, *distrust* people, or *cling* to irrational biases, you hamper creative ideas. To put good ideas to work often requires that you take a chance. Sure, people may laugh. But what do you care if they laugh? One good idea that works the way you hope can silence laughter forever.

Learn to take a risk. Some engineers and scientists tend to be overcautious in everything they do. Caution is an admirable trait worth cultivating. But when you get a good idea, you may find it wise to be incautious for awhile. Few of us can achieve much in our professions without taking an occasional risk in trying out or presenting our ideas.

You're not alert. Good ideas can slide through our minds without ever being recognized—if we're asleep at the switch. Or, after writing out a good idea, we may be too concerned with other matters to recognize the worth of the idea. We may fail to transfer a principle [3] from one field to another, thus losing a useful application of a good idea. Again, we may not detect the differences between causes and effect.

All these roadblocks, and others, can be traced to a lack of alertness for good ideas. To put ideas to work you must always be on the alert for the chance thought that may solve your problem in a flash. Effective creativity can be achieved and ideas made to function if you train your mind to recognize new ideas as soon as they occur. The simplest way to do this is to maintain your alertness at all times.

Convince Others Your Ideas Are Good

Most creative engineers and scientists have trouble convincing others of the worth of their ideas. "This can be very discouraging," says one rocket propellant specialist. "You expend as much energy trying to sell an idea as you do creating it." But since an idea is useless until it's put to work, you will usually have to convince someone that the idea will help him. Here are useful hints on convincing others of the worth of your ideas:

1. *Be ready for the "it won't work boys."* They're almost certain to come up with these three types of negatives:
It's not true; it's impossible; it's useless.
It's probably true, but of no practical use.
It's a good idea, but somebody already thought of it.[3]
Analyze your idea in advance and be ready to counter each of these negatives. Having an answer at hand will give you greater confidence. You will also be better equipped to exert some persuasion. Above all, remember to be polite when arguing for your ideas. Some engineers and scientists tend to be short-tempered when their carefully worked out ideas are criticized. Keep an even temper—you are at a disadvantage when you lose control of your emotions.

2. *Create ideas to sell your ideas.* Don't confine yourself to the good idea alone. Put your imagination to work on ways of getting your idea across. Treat the selling of your idea as a separate problem. Dream up as many schemes as you can to present your idea effectively. Be sure to avoid a stereotyped presentation. This will only waste your time and effort.

3. *Determine the benefits.* Think of the people who will benefit from your idea, the extra profits your organization will derive. Try to win support by showing interested people what they'll gain from your idea. Don't worry about others hogging the credit. If you spearhead a project, nobody will lose sight of your contribution.

4. *Make it easy to say yes.* Waldemar Ayres, Director of Research of the Singer Sewing Machine Company, advises:

> Think through every problem likely to arise in carrying out your proposal. Then provide an acceptable answer to show you've anticipated and planned for every such circumstance. A busy executive has all sorts of worries of his own. If, in order to approve your proposal, he has to stop and solve a problem relating to your baby, the easiest and quickest thing for him to do is to say no.

Charles S. Whiting has nine excellent hints for presenting ideas. Whiting suggests: (1) Select the proper person or group for presentation. (2) Select the right time. (3) Know your audience. (4) Make it factual. (5) Make it clear and simple. (6) Make it well-balanced. (7) Be able to defend it. (8) Emphasize cost and savings or return on investment. (9) Use audio-visual devices or graphics.[5]

Make Creativity a Full-time Interest

Some engineers and scientists can, after long practice, turn their creative energies on and off like an electric light switch. But these men are the exception. For most of us, creative effort is not easy. If we neglect creative activities for long, our skills and interest begin to decrease. So most practicing engineers and scientists find they can produce more ideas when they make creativity a full-time interest. You can acquire and maintain your interest in creativity by using the following 32 springboards to good ideas.

32 Springboards to Good Ideas

Lester R. Bittel, well-known engineering editor, summarized the best advice on getting good ideas. Here's a quick review of his hints,[3] which are based on the procedures of many successful idea-getters:

1. *Find the time of day when you're most creative*—when you're full of drive. That's the time to build a stockpile of good ideas. *Evaluate* ideas when your mind isn't developing creative thoughts.
2. *Build up your idea sources.* Attend technical and scientific meetings, visit important engineering installations, research labora-

tories. Lunch with creative people. Scan a variety of magazines—even those remote from your profession.

3. *State your problem carefully*—don't let the statement suggest the answer. For in suggesting the answer you may close out the opportunity for developing new ideas.

4. *Don't be afraid to work alone.* Creative thinking isn't necessarily a group process. Many a good idea comes from a loner. But have the courage to put your ideas up for criticism. You can use criticism to stimulate your creativity.

5. *Schedule creative thinking periods.* Regularly drill your mind to produce good ideas for your job, your home, your future. Without regular practice your progress will be limited.

6. *Build an idea reservoir.* Ideas seldom fall out of the blue. You must keep flooding your mind with them by studying related information, by experimenting constantly, and by hypothesizing.

7. *Beware of self-satisfaction.* Assume, as Harlow H. Curtice did while president of General Motors, that "anything and everything—product, process, method, procedure, or human relations—can be improved." Self-satisfaction breeds complacency which reduces creative effort.

8. *Organize your approach.* John Arnold, in his MIT creativity course suggested three questions you should ask when replacing a machine or process: (*a*) *Will it do more?* (*b*) *Will it cost less?* (*c*) *Will it be easier to sell to those who must live with it?*

9. *Build big ideas from little ones.* Expand an idea for one portion of your work to the point where it covers all, or a large part of, your task.

10. *Be enthusiastic, confident.* Build faith in your skills by scoring successes on little problems before you tackle the big ones. Remember—your willpower controls your imagination.

11. *Don't worry about waste.* Accept the fact that much of what you produce will be chaff. It's sometimes wasteful to hunt for inspiration. But it's *always* wasteful to wait around for that one big, flawless idea—it may never come.

12. *Be ready for idea flashes.* Relax your mind. Let it wander after a day's work. Try daydreaming, restful music, some strenuous physical activity. Get up an hour earlier and enjoy the dawn. Use the two-day formula. Set your problem aside for a day; hit it hard after a day's rest.

13. *Don't worry about the opinions of others.* Too many people are filled with negatives—they'll give you a thousand reasons "why it won't work." Forget these opinions—go on and make your ideas work.

14. *Keep your eyes open for chances.* Chance favors only the mind that's ready for it. So be alert for any unusual variation, any unexpected happening. Be ready to recognize it and evaluate its chances. The future [5] is open to those who have creative training and know how to use it.

15. *Vary your routines.* Chance favors novel situations. So drive home a different way, choose a different book or magazine to read, try some new way of learning. Shake up your pattern of living. It will help improve your creativity and perspective.

16. *Avoid mind-weakeners.* Liquor, excessive smoking, and pots of coffee usually do not aid creativity. So don't dull your mind—if you do, you're liable to make decisions out of weariness, rather than inspiration.

17. *Avoid fatigue, noise, distractions.* They sap your strength. Keep your body and mind fit—get enough rest, eat moderately, exercise regularly.

18. *Split your problem into pieces*—into pieces that follow a logical sequence. Successful solution of each part of a problem provides an impetus to you to continue working until the entire problem is solved.

19. *Learn to recognize your mistakes.* When you learn to see your mistakes, find out why you made them, and correct the mistakes, you are learning how to think differently—how to get useful ideas.

20. *Beware of vague ideas.* Force yourself to reduce every idea to a specific proposal. This gives your mind a definite problem to solve.

21. *Use the blind-man method.* Shut your eyes—bring your other senses into play. Try doing things by sound, sight, touch. This technique opens new idea channels.

22. *Use the X method.* George B. Dubois, Professor of Mechanical Engineering, Cornell University, suggests that if you've split your problem into pieces and still can't get started, just try the "X" method—go on to any unknown. Solve all the problems that surround the unsolvable problem. Solution of the core problem may then be simple.

23. *Use the change-a-word method.* Dubois' brainchild, too. State

your problem in five or six words (suggestion No. 3). Then substitute another word for any of the five or six without changing the meaning. Take a typical industrial problem: *Provide more electric power.* Now replace *provide* with *buy*—a new thought, and a possible solution. Or for *electric* substitute *d-c, a-c, surge, standby, heating, cooling.* More new ideas.

24. *Try "attribute" listing.* Robert Crawford, University of Nebraska, suggests you list all the attributes of a method. Then think how you can improve them.

25. *Use a checklist.* Alex Osborn calls them idea needlers to help in "piling up alternatives." [6] To find a new way of doing something, go down a list of questions like this: *Adapt?* What else is like it? What other idea does this suggest? Does the past offer a parallel? What could I copy? . . . *Modify?* Change color, meaning, motion, sound, odor, shape? . . . *Magnify?* What can I add? Higher, longer, stronger, bigger, thicker? . . . *Minify?* What can I subtract? Make it lighter, shorter, break it up, omit it? . . . *Substitute?* Another ingredient, material, process, power source? . . . *Rearrange?* Can we interchange components? Use another sequence, layout? Change pace, schedule? . . . *Reverse?* Use opposites? Backwards? Upside down? . . . *Combine?* A blend, an assortment? Can we combine purposes, appeals, goals, ideas?

26. *Try the input-output scheme.* This technique is one used by General Electric in its creative engineering program. Start the solution of your problem by listing the desired input of the new method —what it should do for you. Then list all the things that go into the process that are desirable, necessary, available. Between the two extremes, list the limitations of the "need area."

27. *Sharpen your sense for problems.* Listen to complaints. Jot down your own complaints about the way things run in your department. Ask people outside your immediate work group—accountants, methods men, salesmen—if they see ways you can improve your procedures. These people will usually be glad to tell you what they think.

28. *Develop an idea-conscious mind.* Gather your ideas everywhere. Don't be afraid to associate ideas freely. Let your mind roam from one idea source to another.

29. *Be ready to note down ideas.* Carry a pad and pencil at all times. Ideas are elusive. Unless you make a note of *all* your ideas,

you'll probably remember only the mediocre ones. Good ideas have an exasperating way of disappearing. So use the note method—it pays off.

30. *Set quotas and deadlines.* Force yourself to come up with more ideas—sooner. In one creative training program the men try to develop at least eight workable solutions to every problem. A deadline keeps you from putting things off from day to day.

31. *Use idea banks and idea museums.* That's what Charles Clark, of Ethyl Corporation, calls them. He suggests keeping a file of notes, clippings, pamphlets, etc., even if you can't work on them immediately. As an idea museum, use catalogs, books, reports, and other documents related to your field.

32. *Talk it over.* Sometimes the most obvious method—going to other people for their help—is the simplest way to get ideas. Other people's views will give you a fresh slant on your own problems.[3]

Creativity—Important Today and Tomorrow

Charles S. Whiting, Market Planning Corporation, observes in his fine book *Creative Thinking:*

> As our life becomes more and more complex, and with the advent of automation, more and more routine tasks are turned over to machines. Creative ability will become even more important, for although man has been able to devise electronic calculators or "brains" which can perform amazing feats in terms of memory and repetitive calculations at tremendous speeds, there is no machine which can produce, correctly evaluate and implement a creative new idea. Creative ability is a unique ability of mankind.[5]

Every engineer or scientist who wants to achieve more in his profession can—if he will but improve his creative abilities. Begin now, by using the hints in this chapter, to improve your creativity. Every advance you make will bring you one step closer to an outstanding professional career.

Lastly, must you be unconventional and eccentric to improve your creative powers? Emphatically no. To quote Charles S. Whiting again: "Many of those who have studied creativity have concluded that there is much that can be done to enable an individual to increase or at least better utilize his innate creative ability." Resolve today that you will not be one of those people James Bryce referred to when he wrote "To the vast majority of mankind nothing is more agreeable than to escape the need for mental exertion. . . . To most people nothing is more troublesome than the effort of thinking." Follow, instead, Richard Jefferies' advice "Let me exhort everyone to do their utmost to think outside and beyond our present circle of ideas. For every idea gained is a hundred years of slavery remitted."

REFERENCES

[1] E. J. Tangerman, "Creativity . . . the Facts behind the Fad," *Product Engineering,* July, 1955.

[2] C. D. Tuska, *Inventors and Inventions,* McGraw-Hill Book Company, Inc., New York, 1957.

[3] Lester R. Bittel, "How to Make Good Ideas Come Easy," *Factory Management and Maintenance,* March, 1956.

[4] Alex F. Osborn, *Applied Imagination,* Charles Scribner's Sons, New York, 1957.

[5] Charles S. Whiting, *Creative Thinking,* Reinhold Publishing Corporation, New York, 1958.

[6] Alex F. Osborn, *Your Creative Power,* Dell Publishing Co., Inc., New York, 1961.

HELPFUL READING

Barber, Bernard, "Resistance by Scientists to Scientific Discovery," *Science,* Sept. 1, 1961.

Bender, J. F., *How to Sell Well,* McGraw-Hill Book Company, Inc., New York, 1961.

Bruner, J. S., J. J. Goodnow, and G. A. Austin, *A Study of Thinking,* John Wiley & Sons, Inc., New York, 1956.

Cros, Gamble, Mraz, Whiting, et al., *Imagination—Undeveloped Resource,* Creative Training Associates, New York, 1955.

Curtis, Charles Pelham, and Ferris Greenslet, *Practical Cogitator,* Houghton Mifflin Company, Boston, 1953.

Goldner, B. E., *The Strategy of Creative Thinking,* Prentice-Hall, Inc., Englewood Cliffs, N.J., 1962.

Mason, J. G., *How to Be a More Creative Executive,* McGraw-Hill Book Company, Inc., New York, 1960.

Reif, F., "The Competitive World of the Pure Scientist," *Science,* Dec. 15, 1961.

Spencer, G. J., "Characteristics of the Creative Engineer," *Mechanical Engineering,* February, 1958.

Tangerman, E. J., "Creativity: What Is It?" *Product Engineering,* Dec. 11, 1961.

6

Improve Your Problem-solving Ability

> His [Newton's] peculiar gift was the power of holding continuously in his mind a purely mental problem until he had seen through it. . . .
>
> JOHN MAYNARD KEYNES

In Chapter 5 we discussed the creative approach to solving problems in engineering and science. As Charles S. Whiting says, the creative approach is not the only one, nor is it

> . . . intended to replace the more traditional methods of problem-solving by logic based upon a thorough analysis of the relevant facts. Rather, they [creative-thinking techniques] supplement these methods by providing new ways to develop alternative solutions. In most cases, a combination of problem solving by means of traditional methods such as logic, or the scientific method, and the use of operational techniques [creative thinking], makes the most effective over-all approach.[1]

You have already learned the basic principles of creative thinking. In this chapter you will review and learn how to apply the latest methods for solving everyday problems in engineering and science.

Developing and strengthening your problem-solving ability will equip you to achieve more in every professional activity.

Choose and Use Professional Procedures

Practicing engineers and scientists are fond of analyzing and explaining their work [2] in terms of the four M's:

> *Methods:* Primarily the application of the principles of mathematics, physics, and chemistry to a problem.
>
> *Materials:* The knowledge of the physical and chemical properties of an ever-widening range of materials, both natural and synthetic.
>
> *Money:* Knowledge and experience in estimating and balancing costs to determine the most economical solutions.
>
> *Men:* A knowledge of human experience and human relations to enable the engineer or scientist to work effectively with individuals and organizations.

The four M's enter almost every professional problem met today—they also occur in many scientific problems. John Charles L. Fish, in his book *The Engineering Method,* points out: "The purpose of the scientist is to add to the world's store of systematic knowledge and to discover laws of behavior in nature; the purpose of the engineer is to apply this knowledge to particular situations by deducing specific behavior from the scientists' general principles." [3]

Most professional engineers use, in general, a five-step procedure [2] in solving problems. These steps are: (1) definition of the problem, (2) selection of methods and procedures, (3) collection of data, (4) application of methods and procedures, (5) conclusions and recommendations.

If you're a graduate engineer or scientist, you were trained to use this five-step method, or a similar one. Probably by now you are so familiar with the steps that you hardly think of beginning with a definition of the problem and proceeding to conclusions or data.

Many recent studies show that you can improve your problem-solving ability by updating the procedures you use. Let's see how you can apply the five steps listed above to modern technical and scientific problems involving all or many of the four M's.

Define your problem. Recognize that as you advance in your career the problems you meet will involve much more than simple equations, catalog data, or standardized computation forms. Your problems will become more complex—they will often involve human beings. To date no one has derived a complete equation that will apply to any human being. So it is important that you define your problems in more accurate terms. For "it is only by controlled and organized thinking that the solving of problems can be put upon an efficient, effective basis."[4]

Learn to back off mentally before trying to define your problem. Ask yourself: (*a*) Is this situation confronting me really a problem? (*b*) What is the main problem in the situation? If your answer to (*a*) is yes, you are ready to consider (*b*), the core of the problem. Here is an example of this procedure.

Example. You are an engineer responsible for the design of missile-guidance systems. Your supervisor asks you to study all available guidance systems and report on their relative worth.

Procedure. Review the request by backing off. (*a*) *Is this assignment a problem?* Yes, it is because the data requested are not readily available—they are probably difficult to obtain; you are somewhat uncertain that you can obtain all the needed data; there may be some doubt in your mind as to the reliability of the data you do obtain. Thus, this assignment is a problem because *it presents a difficult question, the solution of which is uncertain, trying, and doubtful.* Real problems are characterized by difficulties in solution, uncertainties as to the outcome, doubts about suitable methods of solution, and perplexing questions as to which is the best procedure. Simple decisions like which of two nearly equal routes to take when driving home or what to have for lunch are not true problems. A true problem presents a difficult situation, the resolving of which will be important to a person or group of persons.

Now that you are convinced that the situation is really a problem, you can apply the second step: (*b*) *What is the main problem?* In the problem given above the main requirement is the *collection of reliable data from all pertinent sources.*

With steps (*a*) and (*b*) completed you are ready to define the problem. Thus, you might define the above problem this way: *Determine all the manufacturers of missile guidance systems. Obtain from each manufacturer complete operating data on each system.*

Note that in defining the problem, you have not stated how the problem is to be solved. This comes later. If you state the intended method of solution as part of the problem definition, two difficulties will immediately arise: (1) The definition of the problem will become cluttered with unrelated facts. (2) The solution chosen may not be the best.

In defining any problem, aim at an accurate, short definition that is general enough to permit a variety of solutions. But don't make the problem so general that no solution is possible. Thus, the problem definition: *Methods of improving internal-combustion engine efficiency* is too general—it could lead to hundreds of solutions, many of which might be useless. Stated another way: *Methods of improving four-cycle diesel engine efficiency,* the problem has greater focus, and the resulting solutions will probably be more useful. Some authorities on problem solving suggest limiting the definition of the problem to twenty words or less; others suggest stating the problem in ten different ways, from which the best statement is selected. Use any of these methods to improve your definition of a problem.

Work at developing your sensitivity for defining problems. Recall John Dewey's famous remark, "A problem well stated is half solved." Ready your mind to recognize problems as early as possible in every situation associated with your profession, personal life, hobbies, and other activities. Early recognition of a problem leads to easier definition, faster and surer solution. Recognizing problems and developing the ability to define these problems in concise, accurate terms are important keys to greater achievements in all branches of engineering and science.

Select solution method. Charles P. Steinmetz said, ". . . the most important part of engineering work—and also of other scientific work—is the determination of the method of attacking the problem, whatever it may be."

You can use a variety of methods to solve a given problem. These methods include theoretical analysis, specific calculation procedures, past experience, logic, creative thinking, and trial and error. In almost every problem there is one most efficient solution method. Form the habit of actively seeking the most efficient solution method for every problem you meet. Finding this method will enable you to solve problems faster with less expenditure of energy. You will achieve more in less time.

Choose the most efficient solution method by listing every method you might possibly use. Study this list, comparing the time and effort required by each method of solution. Recognize that the most common method of solving a specific problem *may* or *may not* be suitable for the situation you are considering. Where a routine calculation procedure will solve a problem (like the design of a beam), don't waste time studying other solution methods, unless some special benefit will be gained. Choose the best method and discard the others.

Nonroutine problems, like the design of an entire manufacturing plant, do not lend themselves to easy choice of a solution method. In problems like this you will probably have to use some trial-and-error techniques, some standard design procedures, plenty of creative thinking, and logic. List these methods and assign one or more of them to specific parts of the problem.

In choosing a solution method feel free to be as creative as you wish. Your own past experience, and that of others, is often helpful in guiding your choice of a solution method. Always aim at the simplest and most direct solution method. Some engineers and scientists enjoy contemplating complex procedures for solving problems. The mental exercise derived from such thoughts is fine, but the efficiency of the solution methods envisioned is questionable. Choosing a simple solution method gets you started on the problem. Even if you must later change your method, getting started early will usually permit you to finish sooner.

As you advance in your career, the problems you must solve become more complex. The best solution methods are no longer simple. Instead of a single, clearly defined problem, you find you have a main problem surrounded by many subproblems. Your best procedure in these circumstances is to choose a solution method for the main problem. Then group the subproblems and handle them as a single problem or a group of similar problems.[5] This procedure will usually speed your work.

Collect data. You will collect data during every step in solving a problem. Thus, in defining the problem you will seek some preliminary facts concerning the actual background of the problem. In selecting a solution method you will collect facts and data related to various alternative methods of solving the problem. But in both these steps your collection of data will be somewhat limited because you are seeking items other than data alone. Set out now, as your third step, to actively collect useful data related to your problem.

Collect data from (1) people, (2) catalogs, (3) handbooks and reference books, (4) business magazines, (5) military and industrial specifications, (6) drawings and blueprints, (7) records of previous projects, etc. Note that where a problem may have a number of acceptable solutions—such as which of several suitable passenger aircraft should be purchased for a given airline service—you will probably have to seek facts and data from many sources—including some not listed above. When the problem is the one-answer or analytical type,[1] the needed data will probably be better organized, more concise, and more readily available.

Form the habit of being constantly on the lookout for new data sources. Be alert when you read, visit other organizations, perform research, or study. Make a note of every new data source you find. Sometimes a ready source of data can mean the difference between a profit and a loss in solving a problem, particularly in smaller organizations. Good sources of reliable data increase your problem-solving capabilities, making you a more valuable engineer or scientist. So take advantage of every opportunity you have to collect data sources for the future problems you are almost certain to meet.

Use judgment when collecting data to solve a specific problem. Too much data can confuse you; too little may lead to an inadequate solution. If you must err in collecting data, err on the excess side. While excess data may tend to confuse you, it is far easier to discard unusable data than it is to find additional data in a hurry. With your data on hand you are ready to apply the needed methods and procedures to solve the problem.

Apply methods and procedures. Solving the problem may be a simple or a complex procedure. Much depends on the nature of the problem. In general, most calculation-type problems having a unique answer are simpler to solve than broad-gage design problems involving a complete plant, product, structure, process, or experiment. Numerical problems having a single suitable answer are frequently solved by computers, reducing the time required for the solution. Today, many complex problems are also solved by computers.

Set up the problem by listing the methods or procedures you will use to solve it. You chose these methods or procedures during the second step in solving the problem, after you had defined it. Where several steps must be used to solve the problem, number each step. Alongside each step list the data, mathematical relations, theoretical principles, intermediate values, and other items or techniques you'll use. Then apply the methods you've chosen. Proceed as rapidly as possible. After finishing a step, try to size it up for correctness. See if the values obtained or chosen are within a logical range. When the problem is being solved by a computer, try to obtain early print-outs of results. By studying these print-outs you can quickly determine if the preliminary results are within a suitable range.

Developing a logical sense for the probable results will help you avoid wasted time and effort. Thus, if your problem involves the effect of a new policy on personnel in your organization, you can sound out some of the people working under the new policy. By listening carefully, by observing facial expressions, and by trying to put yourself in the other person's place, you will probably be able to make an accurate estimate of the policy's effect on most of the people you interview.

Make full use of checklists, dimensional analysis, printed calculation forms, and other aids while solving the problem. Never assume that a problem of any kind *must* be solved without aids that may be available. While a specific aid may not be completely suitable for *your* problem, it may suggest useful solution techniques. And remember—a variety of aids are available to help you, be the problem an analytical one or a human relations type. So use your data, methods, and aids to solve the problem. Once you have the results, you are ready to draw conclusions and, if necessary, make recommendations.

Conclusions and recommendations. Simple analytical problems generally require no conclusion—the numerical result is usually sufficient to provide the information needed. But as your professional responsibilities increase, the number of simple problems you have decreases. Complex problems involving judgment, finances, human relations, and a variety of other factors become part of your regular tasks. These problems will usually have some numerical answers as part of a larger result. You will seldom use the numerical answers alone. Instead, you will have to weave them into your conclusions which will usually be presented in words. Once you've reached your conclusions, you can develop the recommendations.

Here are five tips on preparing your conclusions: (1) State each conclusion accurately and concisely. (2) Be certain that each conclusion is stated as objectively as possible. (3) Check to see that each conclusion relates directly to the problem being solved. (4) Where the results indicate more than one alternative, state each alternative solution in the conclusions. (5) Review your conclusions to see that they are comprehensive without being long-winded.

With your conclusions prepared, you are ready to make your recommendations. Never slight the preparation of conclusions and recommendations. Remember that as a professional engineer or scientist your conclusions and recommendations will be followed by people less familiar with the problem than you are. Clear conclusions and concise recommendations will make each problem you solve a step on the road toward greater professional achievement.

Here are seven tips on preparing recommendations after you've solved a problem. (1) State each recommendation in the active voice. Thus, say *Install a 200-horsepower diesel engine,* instead of *It is recommended that a 200-horsepower diesel engine be installed.* Active-voice recommendations are shorter, more direct—and more likely to be followed. (2) Make each recommendation as specific as possible. Whenever possible state the recommendation in measurable terms like gallons per minute, a crew of three pilots, a 100-horsepower motor, etc. Vague recommendations help no one. If you cannot be specific, don't make a recommendation. (3) Arrange your recommendations in the order of their importance—the most important first, followed by those of lesser importance. (4) Number your recommendations so it is easy for your reader to see how many you have made. (5) Evaluate each recommendation in terms of the problem results—be sure all your calculations, estimates, and conclusions are correct *before* making a recommendation. (6) Where money is involved, take special care to state exact amounts in every recommendation. (7) Don't limit recommendations solely to the problem—where factors related to the main problem are uncovered, make recommendations concerning these, stating their relationship to the main problem.

Check Up on Yourself

Here is a useful checklist to help you solve problems. Read it over before you begin the solution of any major problem. Refer to it at other times when you're seeking ways to improve your overall professional capabilities. This checklist will stimulate your thinking and show you how to achieve more whenever you must solve a problem.

PROBLEM–SOLVING CHECKLIST

Problems and You

1. Build enthusiasm for solving problems; look for new problems you can solve in your career, organizations, and personal life.
2. Face up to the fact that greater professional achievement means more complex problems.
3. Know your problem-solving goals.
4. Seek positive results in every problem you solve, even though your recommendation may discourage a projected undertaking.
5. Train yourself in problem-solving techniques.
6. Make full use of creative thinking procedures whenever you face a problem.
7. Collect and use useful checklists, reminders, calculation forms, and other aids to help you solve problems.

Techniques and Methods

1. Never try to solve a problem until *after* you have defined it.
2. Use as few words as possible to define a problem—and be specific, accurate.
3. Put your problem on paper in words, pictures, or sketches. Putting the problem in any form on paper helps you see the elements of the problem more clearly.
4. Do not suggest the solution in the definition of the problem. This will limit your results.
5. Collect facts after you've defined the problem. Get your facts from as many sources as possible.
6. Use the past to help you solve your problem. Learn how others solved similar problems.
7. Don't overlook theories, procedures, methods, and techniques that will be useful in solving the problem.
8. Spend as much time as possible reviewing your facts, methods, and objectives. After an extended thinking session turn away

from the problem and allow your subconscious to work on the problem.

9. Recognize and allow for the important factors in the problem —time, money, people, materials, and methods.

10. Where your problem is a common one, don't seek complicated, uncommon solutions unless the situation demands it.

11. Never overlook the direct, obvious approach to and solution of your problem. Neither may work, but they will get you started on a suitable solution.

12. Where the direct approach will not solve the problem, take the indirect approach, and do just the opposite of what you did in the direct approach.

13. Never worry when trying to solve a problem. Remember that most problems can be solved if you maintain a neutral, unharried attitude.

14. Develop as many alternative solutions as possible. The more solutions you have, the greater your chances of obtaining the most suitable solution.

15. Take action as soon as you've solved a problem. Where possible check the solution on an experimental basis before putting the solution into full-scale use. Then if your solution is incorrect your error will not be too costly.

16. Evaluate your results as soon as possible after you've taken action. Compare these results with your objectives. Adjust your procedures where necessary.

17. As you see results developing, analyze the eventual outcome. Such analysis will put you ahead of the problem, ready to make any other adjustments that may be needed.

18. Be objective in conclusions and recommendations. Learn to distinguish facts from opinions, correct assumptions from incorrect ones, illogical comparisons, and biased reasoning.

Build your problem-solving ability into a strong, useful tool. It will serve you in every step toward greater achievement in your profession.

REFERENCES

[1] Charles S. Whiting, *Creative Thinking*, Reinhold Publishing Corporation, New York, 1958.

[2] C. C. Williams and E. A. Farber, *Building an Engineering Career,* 3d ed., McGraw-Hill Book Co., Inc., New York, 1957, and Lee H. Johnson, *Engineering: Principles and Problems,* McGraw-Hill Book Company, Inc., New York, 1960.

[3] John Charles Lounsbury Fish, *The Engineering Method,* Stanford University Press, Stanford, Calif., 1950.

[4] Forest C. Dana and Lawrence R. Hillyard, *Engineering Problems Manual,* McGraw-Hill Book Company, Inc., New York, 1958.

[5] Joseph G. Mason, *How to Be a More Creative Executive,* McGraw-Hill Book Company, Inc., New York, 1960.

HELPFUL READING

Bullinger, C. E., *Engineering Economy,* 3d ed., McGraw-Hill Book Company, Inc., New York, 1958.

Clark, C. H., *Brainstorming,* Doubleday & Company, Inc., Garden City, N.Y., 1958.

Crawford, R. P., *How to Get Ideas,* University Associates, Lincoln, Neb., 1950.

———: *The Techniques of Creative Thinking,* Hawthorn Books, Prentice-Hall, Inc., Englewood Cliffs, N.J., 1954.

Dennistoun, W. V. and B. R. Teare, Jr., *Engineering Analysis,* John Wiley & Sons, Inc., 1954.

Keyes, K. S., Jr., *How to Develop Your Thinking Ability,* McGraw-Hill Book Company, Inc., 1950.

Kogan, Zuce, *Essentials in Problem Solving,* Arco Publishing Company, New York, 1956.

Osborn, Alex, *Your Creative Power,* Dell Publishing Co., Inc., New York, 1961.

Von Fange, E. K., *Professional Creativity,* Prentice-Hall, Inc., Englewood Cliffs, N.J., 1959.

7

Strengthen Your Memory

Memory is the diary that we all carry about with us.

The Importance of Being Earnest

The true art of memory is the art of attention.

SAMUEL JOHNSON

Every engineer and scientist headed for greater achievement in his profession can profit from a better memory. As a group engineers and scientists have a greater ability to remember facts than many other professional people. But when it comes to people's names and faces, some engineers and scientists flunk every time—they can't seem to remember a man's name more than five seconds. And faces are worse—meet an engineer at a meeting this morning and the chances are fifty-fifty he won't recognize you in the hall this afternoon.

Are these statements meant as criticisms? No; today's engineers and scientists are intelligent, dedicated, capable men. Sometimes, in their preoccupation with their professions they overlook the advantages of a sound, dependable memory. You can benefit from a better memory no matter how reliable your memory may now be. For few of us, regardless of our professional background, get maximum efficiency from our memories. Let's see how you can improve your memory by using a few simple techniques.

Five Helpful Hints

You can remember more if you *want* to, take *action* to, and *review* important facts regularly. Is a good memory important? It most certainly is. For memory is a vital factor in almost everything you do—decision making, planning, calculating, learning new skills, making judgments, problem solving, reading, listening, etc. And in human relations, an area in which many engineers and scientists can improve, memory is as important as any other skill. How can you remember more? Here are five helpful hints.

1. *Decide that you will remember a given fact, number, face, or name.* The very act of deciding that you *will* remember helps set the information more securely in your mind. A good memory doesn't develop automatically—you must take action to remember better.

2. *Take notes about everything you decide to remember.* Note-taking is for the birds, you say? It reveals you as a scatterbrain? Nothing of the sort; by taking regular notes you will not only remember more, and more accurately, you will also relieve your mind of the chore of keeping a number of unrelated facts on constant tap. Each of us seems to have a limit beyond which we can't efficiently load our mind with miscellaneous facts.

Use some regular way for making your notes. Carry a pocket diary, notebook, file cards, or other suitable record. Classify your notes as you enter them. Typical categories you might use include job, travel, ideas, earnings, home, hobbies, etc. By classifying your ideas when you note them down, it will be easier for you to find them when you need them.

Try note-taking for a few months. The new zip and effectiveness it puts into your memory will make you a firm convert. You'll understand why thousands of engineers, scientists, business leaders, and others make active use of notes to help themselves remember better and more accurately.

3. *Take action to remember.* Listen, observe, think, repeat, speak, take notes—do anything that will reinforce your memory of what is taking place. Too many of us rely on a passive state of mind, *hoping* that we will remember at a later date. You *may* remember and you

may not. Action during an experience helps you recall the incident at a later date.

When taking action: (*a*) Try to use as many methods as possible —notes, sight, hearing, feel. (*b*) Try to develop as much interest as possible in the incident. (*c*) Seek the important aspects of the situation. (*d*) Isolate, if possible, those factors that affect your profession, field, specialty, etc. (*e*) Search for additional information by asking questions, observing special demonstrations, operating equipment, etc.

Taking action in any situation brings you more intimately into the crosscurrents of facts, observations, people. At a later date you will be able to more readily recall what went on. These memories will help you relate one fact to another and one person's statements to those of another. Reacting in a variety of ways builds a reservoir of memories you will be able to recall with greater ease and certainty.

4. *Pinpoint the essence of what you want to remember.* As C. P. Snow, the famous English scientist, said, "Scientists regard it as a major intellectual virtue, to know what not to think about." So concentrate on the meaning and importance of what you want to remember. Thus, you will seldom want or need to remember every statement made in a speech or lecture. But in almost every talk you hear there will be certain facts worth remembering.

To retain the meaning and importance of what you hear, do the following: (*a*) Seek the key facts and define them in terms of their usefulness to you. (*b*) Make notes about facts that will be helpful to your supervisor, associates, or employees. (*c*) Collect meaningful numbers—like production rates, hourly wages, unit costs, etc. Any fact or figure that is useful to you is easier to remember if you search for its meaning and make a sincere effort to use it at a later date.

When storing facts away in terms of their meaning be careful not to mis-remember. Some engineers and scientists shun notes, saying "Oh, I'll remember that." At a later date, when trying to remember, their memory will play quirks. We've all met people who remember a value as half or double the number stated by a speaker. Such errors are usually caused by impressions gained *after* the value was committed to memory. These impressions can lead to a gradual alteration of the remembered value, resulting in an incorrect number being cited. The best cure for this trouble is accurate notes made on the spot when you hear the number stated.

Write your notes as clearly as possible so you can easily read and comprehend them at a later date. Where an unusual value or fact is stated, make an accurate note of it. Then underline, box, or use some other device to indicate that the noted value is accurate. Doing this will prevent you from doubting the accuracy of your own notes at a later date. Where you must take copious notes, try using one of the simpler forms of shorthand. Gregg Notehand [1] is a quick, easy-to-learn, brief writing system based on the simple alphabet of Gregg Shorthand. It will improve and extend your retentive abilities by helping you build effective listening and notemaking habits.

5. *Review remembered facts at regular intervals.* Do this by rereading your notes made at important meetings, lectures, discussions. By reviewing your notes you will reinforce your memory of the facts.

When you do not have notes, recall the situation during a few free moments on your way to work, at lunch, before you go to sleep. Direct your mind to the important aspects of the situation—those items you wish to remember most clearly. If you cannot pinpoint these facts, start by recalling the circumstances of the situation—where it took place, who was there, what was said. Remembering these aspects will help lead you to the important aspects of the situation.

Reviewing facts, situations, incidents, and other items has other advantages. Often, when you mentally relive an incident, you acquire perspective. The passage of time, accompanied by the gathering of new facts, may help you see aspects of the situation that were not apparent at any earlier date. You gain a broader understanding which helps you remember better. Also, you will be better equipped to handle similar situations at a later date. Mentally reviewing important past experiences adds to your professional maturity without giving you gray hair or putting wrinkles in your face. Try it and see.

Use Memory Aids

Dr. Donald A. Laird, and his wife Eleanor, well-known writers on important aspects of business psychology, prepared a list [2] of important items that may have to be remembered in business. This list includes:

Appointments made (with outline of points to consider)
Appointments to be made in the future (medical, dental, etc.)
Calls to make (with outline of points to take up)
Coming events of interest
Contract dates
Dates to follow up propositions
Deadlines (with warning an ample time ahead)
License renewal dates
Meetings coming up
Payments due
Promises you have made
Reports due
Reviews of investments, insurance, etc.

You can probably add many items to this list, keyed to your particular professional activities. Thus, you might, as the Lairds did, expand the list to include nonroutine business activities:

Information you want or need
Letters to write (with points to include)
Messages to deliver
Phone calls to make (with outline of points)
Questions for discussion
People you want to talk to today (and about what)
Agreements, conferences
Anything about which you may have to testify later
Ideas dealing with inventions, discoveries

To keep your personal life free of rancour, there are dates and facts you'll want to remember:

Birth dates and anniversaries (particularly your wedding anniversary)
Errands (particularly for your wife)
Shopping list
Dental record
Medical records
Speeches you've made
Employment record (including dates, duties, salary, etc.)
Papers and articles you've published
Tax records

Where can you make notes to help you remember these and many other facts? Your best bet is to use some type of memory aid. Here are a number that are popular with many engineers and scientists.

1. Buy a pocket-size notebook and use it for *all* facts you want to remember. Enter these facts as they come to mind. Date each page when you begin it. Then all you need do is flip back a few pages to find notes you made earlier in the week. Don't try to classify the notes in any way other than general headings for each note. Thus, you might write: "Call Joe D. Give him facts on new price ($98)."

General notes like these will help you remember more with fewer complications. Keep your notebook simple, and it will work for you. Make it complex, and you'll give it up as a hopeless experiment.

2. Use a pocket-size diary, but get one having large enough pages to permit you to enter comprehensive notes. Diaries having more than one day per page are difficult to write in and worse to read. Should you decide to purchase a diary, do so with the full intention of using it regularly. A diary with empty pages is useless. Enter notes in your diary in the same way suggested above for pocket-size notebooks. If you travel much in your professional activities, invest in a pocket-size diary containing colored maps of the world. These maps are useful on long trips because they help you refresh your memory of geography. Maps are also helpful when you're trying to locate areas of the world in which important events are occurring.

3. Try one of the pocket-type reminder diaries. The design of these varies. One type has dated cards for each month. You slip these in the case, ready for any notes you might wish to make. Alphabetized cards are also included for phone numbers and similar entries. At the end of the month you remove the cards from the case and insert a new set for the next month.

4. If you're adverse to diaries and notebooks, try 3 x 5-inch file cards. Carry these in your pocket and use them for all notes. Some engineers and scientists use tabbed index cards for special entries. When a card is completely filled out, they file it in a desk-top file box for future reference. Try the card system soon—it's useful in your daily professional duties and on trips. While traveling you can use a card to make notes of the name, address, phone number, and other details of each person you call on. Use one card per person. On your

return you can have your secretary file the cards according to some useful scheme.

5. Scraps of paper will serve in place of a notebook, diary, or set of file cards. While notes on scraps of paper are better than no notes at all, the pieces of paper are difficult to manage. You may misplace papers, destroy them by mistake, or otherwise lose track of them. Losing one good note or idea can cause you much trouble. So invest in a systemized way of keeping notes. Your memory power and efficiency will zoom.

6. Use call-up files whenever possible in your professional duties. With a good secretary and a neatly arranged call-up or "tickle" file you can unburden your mind of many details. Take care to instruct your secretary on exactly how you want your call-up file run. Remember—a missed call-up file card can cost you and your organization time, effort, money, reputation. So see that her first task *every* morning is to check the call-up file. Then there's little chance of missing an important item.

7. Make full use of your desk calendar for daily activities. Choose the type of calendar that best fits your professional schedule. Stationery stores have a variety of calendars. Try several until you find that which is most satisfactory.

8. Some engineers and scientists like large wall charts or calendars for making notes. The large easel-mounted pad of note paper is also popular. All these have the advantage of a large work area which is constantly in view. You can lean back in your chair and stare at the notes you've made. Just seeing the notes in front of you will often make you take action to finish the task so you can cross out the reminder.

9. Desk pads large enough to replace the conventional blotter are handy for making notes, performing calculations, or sketching your ideas. One disadvantage is that you may throw away a used page with some valuable notes on it. So train yourself to check a page completely before destroying it.

Check Your Memory Habits

Here's an easy test that helps you check your memory habits. Take this test now and score yourself at the end.

	YES	NO	
1.	______	______	Do you seek an important reason for remembering key facts?
2.	______	______	Do you often review key facts to keep them fresh?
3.	______	______	When you hear a key fact do you immediately seek an imporant reason for remembering it?
4.	______	______	Do you periodically decide you *will* remember better?
5.	______	______	Have you recently made a firm resolve to improve your memory?
6.	______	______	Did you ever analyze the profit value of a better memory?
7.	______	______	Are you persistent in your desire and intention to remember?
8.	______	______	Do you try to understand an item fully before committing it to memory?
9.	______	______	Do you try to remember in the step-by-step sequences followed by a process, procedure, theory, etc.?
10.	______	______	Are you memory conscious—frequently recalling that remembering the importance of your memory will make it work *for* you instead of against you?
11.	______	______	Are you actively seeking the meaning of everything you remember?
12.	______	______	Do you try to overcome self-conscious attitudes (nervousness, bashfulness, etc.) that hamper good memory?
13.	______	______	Do you *plan on remembering* before you begin an important activity?
14.	______	______	Have you recently tried using repetition of facts, names, dates, etc., to improve your memory?
15.	______	______	Do you ever try the *three-times rule* (repeating a fact, word, name, etc.) three times a day to impress it on your memory?
16.	______	______	Have you tried to develop an active interest in the items you wish to remember?
17.	______	______	Do you know how to unclutter your mind of useless facts, numbers, names?

	YES	NO	
18.	______	______	Are you courageous? Do you try to remember as much as you can by working your mind to full capacity?
19.	______	______	Have you tried to relate important facts to sight, sound, touch, use, etc.?
20.	______	______	Do you speak names, facts, numbers to help yourself remember better?
21.	______	______	Have you ever checked yourself to see what you forget most frequently—names, numbers, facts, etc.?
22.	______	______	Are you in the habit of recalling your weak memory areas and trying to strengthen them?

Score yourself as objectively as possible. If you have twenty-two affirmative answers, stop reading and go on to the next chapter. With fifteen or more affirmatives you have your memory under good control. To ensure better control, review the negatives and make a firm pledge to exert better control in the future. With ten to fifteen affirmatives you should give your memory some active attention immediately. Less than ten affirmatives calls for some intensive action. Restudy the present chapter and secure some of the books listed below.[3–5] Your memory can be improved by systematic study and effort.

Resolve today to make your mind your best assistant, instead of your worst helper. No matter what your duties in your profession, you can profit measurably from a better memory. And all you need do to take the first step toward a better memory is to resolve to remember more. Try it and see. You will learn, as Thomas De Quincey did, that "it is notorious that the memory strengthens as you lay burdens upon it, and becomes trustworthy as you trust it."

REFERENCES

[1] Louis A. Leslie, Charles E. Zoubek, and James Deese, *Gregg Notehand,* McGraw-Hill Book Company, Inc., New York, 1960.

[2] Donald A. Laird and Eleanor C. Laird, *Techniques for Efficient Remembering,* McGraw-Hill Book Company, Inc., New York, 1960.

[3] Joyce Brothers, *10 Days to a Successful Memory,* Prentice-Hall, Inc., Englewood Cliffs, N.J., 1957.

[4] James E. Deese, *The Psychology of Learning,* McGraw-Hill Book Company, Inc., New York, 1958.

[5] Frederic C. Bartlett, *Thinking*, Basic Books, Inc., Publishers, New York, 1958.

HELPFUL READING

Bartlett, F. C., *Thinking*, Basic Books, Inc., Publishers, New York, 1958.

Barzun, Jacques, *The Modern Researcher*, Harcourt, Brace and Company, Inc., New York, 1957.

Byrne, Brendan, *Three Weeks to a Better Memory*, Bantam Books, Inc., New York, 1956.

Chase, Stuart, *Guides to Straight Thinking*, Harper & Brothers, New York, 1956.

Deese, James E., *The Psychology of Learning*, McGraw-Hill Book Company, Inc., New York, 1958.

Dimnet, Ernest, *The Art of Thinking*, Fawcett World Library, New York, 1959.

Hunter, Ian, *Memory: Facts and Fallacies*, Penguin Books, Inc., Baltimore, 1958.

Nichols, R. G., and Leonard A. Stevens, *Are You Listening?*, McGraw-Hill Book Company, Inc., New York, 1957.

8

Learn to Plan Better

Even a poor plan is better than no plan at all.
ANONYMOUS

It is a bad plan that admits of no modification.
PUBLILIUS SYRUS

The ability to plan well in your professional and personal life can mean the difference between outstanding success and mediocre achievement. Planning—*which is the development of a course of action to accomplish a definite objective*—can be the guiding force that urges you to achieve more than others having greater natural talents. Plans lead to decisions; decisions to action; action to results. The results may not be exactly what you want or expect, but with results you have something to improve. You can't have achievement without results of some kind.

What Planning Ability Can Do for You

The ability to plan can (1) make future work more successful, easier; (2) help you unclutter your workday and personal life; (3) keep you ahead of your daily problems (and who hasn't daily problems?); (4) force you to think your problems through; and (5) help you achieve more in your profession.

Some engineers and scientists scoff at the idea of planning. They'll cross that bridge when they come to it. While this attitude may appear to pay off for some people, *the average engineer or scientist must plan, and plan well, if he is to move ahead in his profession.* There is no substitute for planning. Even luck and the breaks seem to favor the man who plans.

Planning can cover the full range of time available to any man. You can plan what you'll do in the next few minutes, hours, days, weeks, months, or years. If you're thirty today, you may be planning what you'll be doing when you're sixty-five. Or at sixty-five you may be planning what you'll be doing at seventy-five. Planning is always related to some future timing, even though the exact future date is as yet unselected. "Cross-the-bridge" engineers and scientists overlook one important aspect of planning: *Well-made plans give us some control of our future actions.* As controls they may be the factors that lead us to outstanding professional and personal achievements. Plans relate our goals to our resources, schedule, action, and follow-up.

How to Plan Better

The objective of any plan is a decision. Once you make a decision, you can take action to achieve results. So in any planning we must recognize that plans are of little value without resulting decisions. And decisions accomplish little without action.

Here are nine steps in effective planning. Follow these steps the next time you must plan anything in your professional or personal life.

1. *Choose a specific goal.* Avoid planning until you know what you want to accomplish. For without a specific goal you will be lost as to how and in what direction to proceed.

2. *Name your goal.* Write it out in as concise form as possible. Writing out your goal will help you understand it more clearly. Any extremely difficult aspects of the goal will become more apparent, and you will be better able to prepare for them.

3. *List the alternative means to achieve your goal.* Almost every goal can be achieved in more than one way. Knowing each alternative will help you choose the course of action best suited to your

particular situation. Also, should you begin your action along one course and find that the results are not what you're seeking, you can revert to one of the other alternative courses. Thus, listing the alternative means to achieve your goal can be a big help to turning your plans into accomplished objectives sooner.

4. *Study each alternative to see if it can be improved, simplified, or made more effective.* Arbitrarily listing alternative courses does not mean that you'll automatically achieve your goal. Each alternative needs study. Improving an alternative *before* you begin to take action can simplify your task considerably.

5. *Itemize on paper each step you must take in each alternative.* Doing this brings your judgment and experience into play. Thus, you may find that one alternative requires twenty steps; another ten. This doesn't mean that the alternative with the fewest steps is best—each of the fewer steps may be far more difficult than the many steps.

6. *Evaluate each alternative in terms of what may be ahead.* One big secret of effective planners is their ability to anticipate the problems they may meet. By anticipating these problems the planner prepares himself to solve them—if and when they occur.

7. *Plan your actions—know what you will do or say to people.* Mentally review whom you'll meet. If you know these people personally, imagine what action or steps they will take. If you don't know the people, try to anticipate their actions by reviewing their job interests and responsibilities. Thus, you can prepare yourself to meet and work with other engineers, scientists, accountants, managers, etc. To plan well you must know how to work effectively with all kinds of people. See Chapters 13, 15, 16, and 27 for helpful hints.

8. *Choose an attainable time schedule for achieving your goal.* Without a time schedule your plans are static, unproductive. If you are a slow starter by nature, or if you tend to throttle back your efforts toward the middle of a planned activity, shorten your schedule somewhat. This reduced time interval will spur you to earlier achievement of your goal. Knowing you have a shorter period in which to complete your task helps you overcome your inertia or slackening of interest.

9. *Start your task on time.* Make every effort to complete it on schedule. Use all the experience and judgment you can muster to

successfully fulfill your plan. Remember: Good plans are next to worthless without suitable action to produce results.

Evaluate Your Results

Every plan on which you take action will produce results of some kind. These results may be good, poor, or bad, depending on many factors. Even bad results, however, are worth having. For with a poor or bad outcome to a plan you have something on which to build. And the experience you gained while achieving these results will help you improve the outcome of your next attempt. Thus, if you write a report and find on completion that it isn't too good, you can rewrite it. Every time you rewrite the report you acquire more writing experience. If you're alert and try to benefit from experience, you will soon learn how to prepare an effective report that will be helpful to your superiors.

Use high standards when evaluating the results of your plans. Don't be satisfied with half-finished tasks—complete the work as planned. Then you'll have a better chance to evaluate both the results and your own performance. Remember: Making and carrying out a plan provides you with valuable experience. Every plan that is properly executed increases your ability to carry out a responsible task on schedule with the opportunity to evaluate the worth of your performance.

Use creative thinking (Chapter 5) to improve your approach to evaluating results. By constantly striving to improve and refine your results you can soon reach a state where you can predict, with fair accuracy, the outcome of a specific plan. When you reach this level of ability you can truly say that your skill in planning is making a definite contribution toward your professional achievements.

Where to Use Plans

Plans are important in two areas of your life—your professional activities and your personal pursuits. Since both are related, planning can help you accomplish more with less effort.

Think of what parts of your life can benefit from good planning. One simple way to begin learning how to plan while at the same time improving your professional abilities is in the area of new knowledge. Analyze your daily routine. Try to find half an hour during which you can study to improve your skills by acquiring knowledge of new developments in your field. Then lay out a simple plan to acquire a working knowledge of a new professional subject by a certain date. Plan what books you'll read, which courses you'll take. Then put your plan into effect and begin working to achieve your goal by the date you've set.

Apply the methods you learned in this chapter to your professional duties. Instead of following an unplanned daily routine, try planning your work for the next few days. You'll have interruptions, to be sure, but a planned workday will produce far better results than a day in which you allow interruptions of all kinds. See Chapter 4 for some useful hints on building efficiency into your workday by using effective planning.

Develop long-range plans for your career. You can hardly expect to achieve more in your chosen field if you don't have a specific goal. As you saw earlier in this chapter, the quickest way to achieve a desired objective is by forming a step-by-step plan to guide your efforts. Long-range career goals are probably amongst the most important objectives you set for yourself in life.

To formulate career goals, (1) decide what you'd most like to be doing in your profession a specified number of years from now; (2) check the probable financial return this activity would earn for you; (3) relate this return to your expected family responsibilities at the future time; (4) if the financial return is satisfactory, begin planning immediately how you can move from your present position to the desired one; (5) if the financial return will be insufficient, investigate means by which you can accumulate sufficient capital to permit you to work at your desired activity.

Don't be hasty with long-range career plans. Spend a year or more investigating professional opportunities, analyzing your true interests, and projecting the returns you'll receive. Remember: You can begin long-range plans at any age. Any number of engineers have changed their occupations in mid-career and later. Thus, there are engineers and scientists you've left industry to take up teaching careers. Likewise, there are many teachers in our colleges and universities who've left the campus to enter industry. Engineers and scientists leaving industry for teaching often explain their move by saying, in part, "I believe I'll be able to contribute more to my field if I teach. Not only will I be able to guide young people entering the field, I'll also be able to share my industrial experiences with them." The teacher going into industry often explains his move by saying "I want to apply my theories; industry will give me a chance to do so. I'll come back a better teacher."

When making long-range career plans, be sure to recognize the importance of professional achievements and contributions. Young engineers, and some young scientists, tend to scoff at the thought of doing anything but making money. While earning a living is essential, and cannot be overlooked, true joy comes to a man when he earns his livelihood while performing a task he believes is important because it contributes something to mankind. Seek out such a career goal and you will achieve more in your field. This is a worthy objective for any man. As George Bernard Shaw so accurately said, "I don't believe in circumstances. The people who get on in this world are the people who get up and look for the circumstances they want, and if they can't find them, make them." Long-range career planning will help you to either look for or make circumstances. Why not begin today?

Plan Personal Pursuits

Use planning in your personal life to build a fund for your retirement, to direct your children's interests toward a richer life, to improve your health, to provide vacations that will be refreshing for everyone in your family. Good planning can open new areas of happiness and enjoyment for yourself and every member of your family. So begin now to see how you can expand your personal life to provide more for yourself, your wife, and children.

Plans Can Pay Off

If you've wondered why you haven't achieved more in your profession, the reason may be because you've neglected to plan for the future. You can readily change your planning habits by applying the many hints in this chapter. While making plans won't assure you of greater achievements, the time and effort you spend analyzing your present situation could bring you a major return in self-knowledge. Knowing more about yourself and your ambitions can direct you to worthwhile goals. Then adequate planning can provide a chart for steering toward those goals. As one sage wrote, "Nothing is so empty as a day without a plan." The same applies to your career, personal life, and profession.

HELPFUL READING

Haldane, Bernard, *How to Make a Habit of Success,* Prentice-Hall, Inc., Englewood Cliffs, N.J., 1960.

Landy, T. M., *Production Planning and Control,* McGraw-Hill Book Company, Inc., New York, 1950.

Magee, J. F., *Production Planning and Inventory Control,* McGraw-Hill Book Company, New York, 1958.

Maynard, H. B. (ed.), *Top Management Handbook,* McGraw-Hill Book Company, Inc., New York, 1960.

Uris, Auren, *The Efficient Executive,* McGraw-Hill Book Company, Inc., New York, 1957.

9

You Can Write Better

> A man may write himself out of reputation when nobody else can do it.
>
> THOMAS PAINE

Probably more criticism has been leveled at engineers and scientists for their poor writing ability than for any other shortcomings they may have. Surveys of all kinds label the engineer as an "atrocious writer," "insensitive to management's communications needs," etc. The criticisms and the surveys are not always accurate. Some engineers and scientists *are* capable writers. But many of today's engineers and scientists are unwilling to expend the time and energy needed to improve their writing. You *can* become a better writer, if you try. This chapter shows you how.

What's Wrong

Poor writing in the technical field can usually be traced to one major fault—trying to give too much detail and too many facts to the reader. Many engineers and scientists hide their main findings or conclusions under layer after layer of abstruse technical information. As a result the reader becomes mired in a mass of information he can't use, hardly understands, and dislikes reading. The reader tosses the written piece aside. He loses because he obtains only a small part of the information he sought. The author of the piece also loses because only a portion of his message is understood by his readers.

You can be the poorest grammarian alive, your spelling can be awful, and your punctuation worse—these faults hardly matter if you know how to organize your written material for easy understanding and use. Grammar you can learn, or if you're high enough in your organization, you can have a competent grammarian correct your errors. To spell correctly all you need is a dictionary. Punctuation can be improved—there are many fine books available on the subject.

But organization of technical and scientific material takes skill. It requires (1) a knowledge of the subject, (2) recognition of the reader's interests and needs, and (3) grouping of information and facts to best serve the reader. You already have a knowledge of your field. With a little study you can learn to recognize your reader's interests and needs. Combining these two—knowledge and recognition—you can group your data to best serve the reader. Let's see how.

Name your reader. Who is he? Is he a graduate engineer, a scientist, or a member of management? Write his title at the top of a sheet of paper.

President or chairman
Vice-president of manufacturing
Chief engineer
Project engineer
Design engineer
Chief scientist
Research scientist
Technicians
Military personnel

Once you've named your reader, decide what type of information he needs about the subject under consideration. Thus, suppose you're writing a report for the president of your organization. As the top management official he will be interested in finance, profits, sales, costs, returns, etc. He will have less interest in engineering designs, manufacturing procedures, and technical problems. Figure

9.1 shows how his interests in management versus technical matters will be divided.

Write the probable interests alongside the man's title. Thus, you might write:

"President: Costs; sales (in units); profits; finance"

Once you've named your man and his interests, you are ready to organize your material. But before we discuss the steps you can take to organize your material, let's look at what you already know.

You've named your man and his interests. Note that you can use this method for any written piece—report, letter, article, technical paper, etc. With some of these written pieces you may have hundreds or thousands of readers. Certainly, you can't name each reader, but you can name a class of readers. If you're writing an article for *Product Engineering* magazine, you might name your reader as mechanical product designer. His interests could be classified as product durability, appearance, engineering design procedures, etc.

When writing a letter, the title of the recipient will often give you all you need know. Thus, chief scientist, electronic countermeasures, tells you enough about the man to organize your material so it will be most beneficial to him. You can now see that this technique can be used for every item you must write.

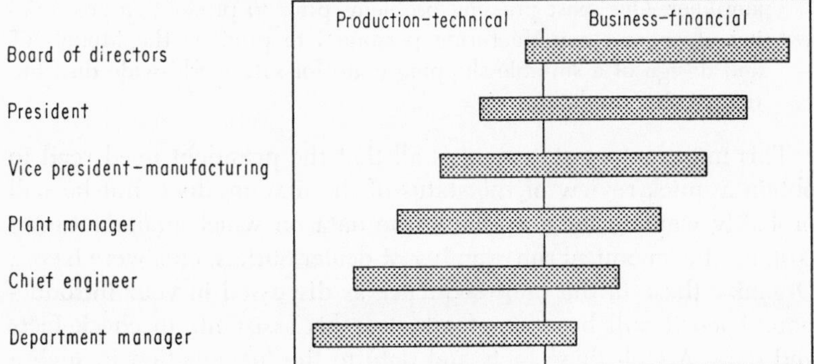

FIG. 9.1. Shift of interest in report content in a typical industrial firm. (*W. A. Ayres and Factory Management and Maintenance*)

Organize your facts. Select information your reader needs. Assume you're preparing a report on a new transistorized amplifier for the president of your organization. This amplifier will be manufactured by your firm. Your report will recommend the size of production runs, materials to be used, etc. But before the president can approve these recommendations, he must have pertinent financial data. These data must be presented first. For if they are favorable, it will be easier to approve the manufacturing methods to be used.

Jot down all pertinent financial data you have—costs, capacities, sales potentials, etc. Your notes might look like this:

> Amplifier manufacturing cost: $2.18; most economical run: 10,000
> Probable list price: $10.95; dealer outlets: 80,000 (estimated)
> Sales potential: 45,000 per year; saving over present product: $1.12/unit
> Problems: training manufacturing personnel; shipping-crate design

With these facts available you are ready to begin your report. Though we've called this item a report, it needn't be in bound form —it might be a letter, memo, etc. Write out what you think the president will be most interested in. You might write it like this:

> The new Model AT transistorized amplifier will cost $2.18 per unit to manufacture. With a probable list price of $10.95 the most economical run will be 10,000 units. Estimated number of dealer outlets for this unit is 80,000; annual sales potential is 45,000 units per year for the first three years. The Model AT can be manufactured at $1.12 less per unit than our present three-stage vacuum-tube amplifier. Our most pressing problems prior to production are training of present manufacturing personnel to produce the Model AT and design of a suitable shipping crate for safe world-wide distribution of the amplifier.

This introductory statement is all that the president need read to obtain a quick review of the status of the new product. But he will probably want to have available the data on which manufacturing cost, most economical run, number of dealer outlets, etc., were based. Organize these in the same sequence as discussed in your introduction. Then it will be easier for him or his assistants to check facts and data. Attach these facts and data to the introduction in such a way that they are obviously supplemental data. You can do this by labeling, leaving a large space, or binding the information separately.

Use any of these, or other techniques—the important fact to keep in mind is that your reader should see at a glance what you have singled out as important and what you are using to verify and support this information.

Why is this method of presenting written information effective? Because, as Figure 9.2 shows, you save your reader much work. No reader today, regardless of his position in life, has the time to make a detailed study of your written piece to trace every step you took. Instead, he wants facts, results, conclusions, or predictions first. With this information in mind he can then proceed to the details, if he wishes or needs to.

Another example. Read the following introduction to an article from *Science.* Note how it summarizes the author's main thought.

> The "pure scientist" is likely to be pictured as a person who devotes himself to study of natural phenomena without regard to their possible practical or technological applications. Motivated by intellectual curiosity and immersed in his abstract work, he tends to be oblivious of the more mundane concerns of ordinary men. Although a few older scientists have become active in public affairs in recent years, the large majority who remain at work in their university laboratories lead peaceful lives, aloof from the competitive business practices or political manipulations of the outside world.
>
> There is some truth in this stereotyped portrait. But if a young student took its apparent serenity too seriously, he would be forced

FIG. 9.2. Comparison between study and report sequence. (*Tom Johnson and American Machinist*)

> to revise his perspective very early in his scientific career. The work situation of the scientist is not just a quiet haven for scholarly activity, ideally suited to those of introverted temperament. The pure scientist, like the businessman or lawyer, works in a social setting, and like them, he is subject to appreciable social and competitive pressures.[1]

This fine article proceeds to prove the assertions made by the author in his introduction. To prove his claims the author cites a number of examples of competitive situations in the world of science. These examples strengthen the claims he made in the introduction to his article. Yet, if you didn't have time to read these examples, the introduction quoted above would be sufficient to alert you to the fact that the working world of the pure scientist can be extremely competitive. Besides citing examples the author also discusses prestige, success, publication of scientific papers and letters, and conflicting values. These discussions amplify his main contention and serve as evidence to prove his claims.

Compare this article with the report you studied earlier in this chapter. The author of the article begins by presenting his findings—the scientist is subject to appreciable social and competitive pressure. As a scientist or manager of scientists this can be an extremely useful fact to you. Yet you acquired this information after reading only a few sentences. The author was able to increase your store of knowledge with a minimum effort on your part. Should you care to do so, you can read the cited examples and other supporting facts.

In the report the introduction summarizes financial data. The *Science* article summarizes the main finding of its author. Supporting evidence follows the main findings in both instances.

Use your head. "Whatever we conceive well we express clearly," said Boileau, the French writer. Most engineers and scientists have a good, clear concept of their work. Where they sometimes fail is in *thinking through* the best way to express clearly their findings. You can double or triple the effectiveness of your writing by carefully isolating the facts most useful to your reader and presenting these first.

Here are some helpful guides to selecting and presenting facts: (1) Every time you find it necessary to describe either a simple or intricate process, experiment, procedure, etc., extract from this the main facts. Set these aside for presentation in your introduction.

Thus, for a process you might isolate the production rate, units per hour, and unit cost. For an experiment you might isolate its ease of performance and the reliability of its results. In discussing a procedure you could cite its usefulness in detecting certain important information. (2) Present results, findings, conclusions first. Follow these with your evidence. (3) Key your results, findings, and conclusions to your probable reader. Emphasize items of greatest interest to him. (4) Be brief—don't try to impress your reader with the depth of your knowledge. In general, the briefer your written items are, the more forceful they will be, if your information is carefully chosen. Reading is work—make it easier by being concise and accurate.

You now have a good concept of one useful way to organize your information. So let's look at some specific hints you can apply in various kinds of writing. These hints will help you fit your facts into an acceptable and convincing pattern.

Business letters. Professor Robert L. Shurter,[2] of Case Institute of Technology, has many valuable hints for business letter writers. Here are five of these hints you can use in the next letter you write:

> 1. Use the *you* attitude—take the reader's point of view in every letter you write. Instead of telling the reader what he should do for you, tell him what he can do to help himself. *Example:* (Poor) Please answer as soon as possible. It's essential that we have an immediate reply so we can make our plans. *Example:* (Improved) Your opinions about past programs and our tentative plans for this year will help us to serve your needs. If you will fill out and return the questionnaire as soon as possible, we can help you promptly.
>
> 2. Keep your reader's interests and desires in mind—be sincere—don't be superficially cordial. Unless you use the you attitude sincerely and in good faith, you pervert its intent and defeat its purpose. Sam Johnson knew this when he said "In a man's letters his soul lies naked." Properly used, the you attitude tells the reader in an honest, tactful, truthful manner the benefits he obtains from an action or an attitude implicit in your letter.
>
> 3. Adapt your tone to your reader—talk the reader's language. Keep in mind that your reader is more than a name—he's a human being. Be sure you know your reader's business, position, reason for writing, and the answer he expects. Thus, if you're writing a letter to a jet-engine mechanic, don't discuss the advanced mathematics of blade-flutter analysis. The mechanic probably won't understand

you. Concentrate instead on the practical aspects of jet-engine maintenance with which the mechanic is concerned and familiar.

4. Begin your letter with a short paragraph—make it say something. As a general rule, never use more than two or three sentences in your first paragraph; if you can use fewer, so much the better. Two things should appear in your first paragraph: (a) date of the letter being answered or similar details to give continuity to the correspondence, (b) a statement of what this letter is about, unless it is a sales letter in which the first paragraph is designed to attract attention.

5. Follow this principle in the last paragraph of your letter: Stop when the message is complete. The function of the last paragraph of every letter is to make it as easy as possible for your reader to take an action or accept a point of view that you want him to take. Use the you attitude to show your reader how easily he may do that which will benefit him. *Examples:*

> "May I have an interview with you at your convenience? You can reach me at my home address or FA-5-4289."
>
> "If we can help you in any way, please let us know."
>
> "Mail us your check today and your order will arrive on Thursday, May 15th."

Memos are important. Next to letters you probably write more memos than any other type of written communication. Memos are important in all organizations. Study the memos you write, and you'll find that they (*a*) supply information, (*b*) ask for information, or (*c*) establish policies or procedures for the organization.

To obtain best results from your memos, do the following: (1) Limit each memo to only one topic; if you have more than one topic to discuss, use a separate memo for each. (2) Be brief—give the facts and stop. (3) Eliminate long introductions, lengthy descriptions of the background of the subject. (4) Put important facts first; follow with other facts in an order of decreasing importance. (5) Number important items; underline key words. (6) Use the memo form provided by your organization. See that its various elements ("to," "from," "date," "subject," "location," etc.) are completely filled out. Neglecting these items can lead to confusion, wasted effort, and loss of time.

Professional reports. Letters and memos are your main day-to-day writing task. Reports, though written less often, may be far more important than routine correspondence. For a good report can mean the difference between acceptance or rejection of your professional recommendations. Good reports, well-written and carefully prepared, can also mean the difference between outstanding professional achievement and mediocre accomplishments. Since the report gives you a major opportunity to contribute to your field, you are wise to make every effort to turn out good reports. Here are six helpful hints you can use when writing any report—long or short.

(1) *Start your report with a summary of the main findings, conclusions, or recommendations.* Limit your summary to 100–200 words. Direct it at the probable recipient of your report by stressing those topics which interest him. See Figure 9.1 to determine the shift of interest from one professional position to another. (2) *In the summary try to answer the questions your reader is likely to ask*—profits, costs, labor savings, reliability, weight, size, performance, efficiency, advantages, disadvantages, competitive aspects, etc. (3) *Arrange your report in accord with the recommendations of your organization.* One popular arrangement is: title; table of contents; abstract, summary, or conclusions; preface; body; appendixes. (4) *Plan your report for easy use.* Choose informative headings, underline key facts, and use colored section tabs and other devices to highlight important information. (5) *Deliver your report in person, if possible.* Tell your supervisor in concise terms what you found. Subordinate details—emphasize your main findings. (6) *Write your report with these aims*[3]*:* (*a*) Be specific; (*b*) aim for short sentences; (*c*) beware of qualifying clauses; (*d*) keep the ideas per sentence low; (*e*) stick to direct statements—active verbs; (*f*) get rid of say-nothing words.

Technical and scientific articles. You can derive many benefits from writing good articles related to your professional activities. These benefits include professional recognition of you by the people in your field, a contribution of knowledge to your field, prestige for your organization, improvement of your knowledge of the subject, and the possibility of extra income. Here are nine steps to follow when writing any technical or scientific article.

> (1) Choose the type of article you intend to write.[4] (2) Decide which magazine would be most likely to publish your article. (3) Study at least six recent issues of the magazine to determine the best method of presentation for your material. (4) Prepare an outline of the article. (5) Collect illustrations and tables. (6) With the outline and illustrations ready, write the article. Use your study

of the magazine as a guide to the number of words, illustrations, and tables you should have in your article. As a rough estimate, plan on using about two illustrations per page for articles of four or less published pages. The usual magazine page contains about 1,000 words when no illustrations are used; 700 words when the illustrations occupy about one-third of the printed page. (7) Write quickly—try to finish the text as soon as possible. (8) Concentrate on technical or scientific facts while you write. Get your thoughts on paper. You can improve the grammar later. (9) Have your article typed double-spaced and submit it, along with the illustrations, to the magazine you chose earlier.

Professional papers. Technical and scientific papers comprise the permanent literature of almost all learned fields. Engineering, scientific, and technical societies publish papers for the benefit of their membership and their field. A well-written learned paper can bring you many of the same benefits as an article, plus some others. In preparing and publishing a paper, you register your name among the most outstanding men in your field. For learned papers form a permanent body of reference material for a given field. Here are twelve hints you can follow when writing technical or scientific papers. Try them when you write your next paper—they really work.

(1) Understand what a paper is. *A technical or scientific paper is a comprehensive presentation of the results of study, research, experimentation, testing or experience in a given area of technology or science.*[4] The subject of a paper can vary from the most advanced theories of a field to the practical everyday aspects of a job. (2) Decide which learned society will be most likely to publish your paper. Usually this is the society of which you are a member. But note that you need not be a member of a learned society to have your paper published. (3) Determine the maximum manuscript length permitted by the society—thus the American Rocket Society * specifies a typical maximum length of twenty-two double-spaced typed pages and twelve illustrations. This information, and other rules to be followed, is available from the editorial secretary of the society to which you intend to submit your paper. (4) Prepare an outline of your paper. A typical sequence recommended by one professional society is: title; author, job title, organization, abstract, body of paper, appendixes; acknowledgements, bibliography, tabulations, il-

* Now named American Institute of Aeronautics and Astronautics.

lustration captions, illustrations. (5) Write an abstract of the paper —try to limit this abstract to 100–200 words. (6) Begin writing the body of the paper as soon as the abstract is finished. *Start with familiar facts and lead the reader to those parts of the subject that are new or unfamiliar to him.* (7) Keep your specific audience clearly in mind [3] while writing the paper. Remember that almost every member of a learned society is a college or university graduate. But this doesn't mean your writing should be complex, pedantic, or abstruse. Be clear, concise, and simple. (8) Use subheads to indicate the major divisions of the paper body. (9) Concentrate on presenting your data as concisely as possible. There is a trend today toward shorter, less complex technical and scientific papers. (10) Be as objective as possible while you write. (11) Use an appendix for derivations of equations, long tabulations, detailed descriptions of apparatus, etc. (12) Include acknowledgements and a bibliography where such sections are necessary.

General Writing Hints

You can improve every piece of writing you do by keeping some general writing hints in mind. Here are twenty hints given by two specialists in the art of clear writing. The first, Robert L. Shurter, of Case Institute of Technology, suggests the following:

> (1) Keep your sentences short. (2) Put your qualifying ideas in separate sentences. (3) Use paragraphs to break your text into readable units. (4) Use the active voice of verbs wherever possible. (5) Make your verbs carry their share. (6) Prefer the short word and the specific expression. (7) See if *which* and *that* clauses can be expressed more concisely. (8) Rearrange your sentences to make them more direct. (9) Show the relative importance of ideas within the sentence. (10) Avoid business jargon.[2]

Chester R. Anderson, of the University of Illinois, suggests that if you want to achieve a readable style of writing, remember to:

> (1) Use short paragraphs—they make for readability. (2) Avoid long and involved sentences—an 18 to 20 word *average* is far better than 40 to 60. (3) Watch conditional clauses and prepositional phrases—the more you have, the harder the reading. (4) Use topic sentences—they provide good road maps for paragraphs. (5) Watch transitions—be sure that one sentence follows

another—use transitional arrows to point both forward and backward. (6) Vary sentence structure—don't write all "subject-verb" sentences, or start them with such dead words as *the, it,* or *there.* (7) Watch repetition of the same word or phrase. (8) Break down figures so that they are understandable—and don't use too many in the narrative. (9) Make use of display, headings, enumeration, parallel construction, etc. (10) Leave plenty of white space.[5]

Every engineer and scientist *can* improve his writing ability. All you need do is to continually seek to make each written piece better than the last. Remember what Epictetus said: If you wish to be a writer, write. And as a well-known engineering editor remarked, "Other things being equal, skill with words will add between $20,000 and $100,000 to an engineer's life-time earnings." Today many firms pay their engineers and scientists an honorarium for the preparation of worthy articles or papers. Narda Microwave Corporation awards a bonus to its engineers for an acceptable technical paper *before* the paper has been accepted for publication. So you see, the rewards for good writing are well worth the effort.

REFERENCES

[1] F. Reif, "The Competitive World of the Pure Scientist," *Science,* 15 December, 1961.

[2] Robert L. Shurter, *Written Communication in Business,* McGraw-Hill Book Company, Inc., New York, 1957.

[3] Norman G. Shidle, *Clear Writing for Easy Reading,* McGraw-Hill Book Company, Inc., New York, 1951.

[4] Tyler G. Hicks, *Writing for Engineering and Science,* McGraw-Hill Book Company, Inc., New York, 1961.

[5] Chester R. Anderson, A. G. Saunders, and F. W. Weeks, *Business Reports,* 3d ed., McGraw-Hill Book Company, Inc., New York, 1957.

HELPFUL READING

Buckley, E. A., *How to Write Better Business Letters,* 4th ed., McGraw-Hill Book Company, Inc., New York, 1957.

Flesch, R., *How to Make Sense,* Harper & Brothers, New York, 1954.

Harwell, G. C., *Technical Communication,* The Macmillan Company, New York, 1960.

Hicks, Tyler G., *Successful Technical Writing,* McGraw-Hill Book Company, Inc., New York, 1959.

Linton, C. D., *How to Write Reports,* Harper & Brothers, New York, 1954.

Mandel, Siegfried, *Writing in Industry,* Polytechnic Press, New York, 1960.

Murphy, Dennis, *Better Business Communication,* McGraw-Hill Book Company, Inc., New York, 1957.

Rathbone, R. B., and James B. Stone, *A Writer's Guide for Engineers and Scientists,* Prentice-Hall, Inc., Englewood Cliffs, N.J., 1961.

Raymond, H. J., *The Research Paper,* Henry Holt and Company, Inc., New York, 1957.

Robertson, H. O., and Vernal Carmichael, *Business Letter English,* 2d ed., McGraw-Hill Book Company, Inc., New York, 1957.

Shaffer, V., and Harry Shaw, *McGraw-Hill Handbook of English,* 2d ed., McGraw-Hill Book Company, Inc., New York, 1960.

Shurter, Robert L., *Effective Letters in Business,* 2d ed., McGraw-Hill Book Company, Inc., New York, 1954.

Ulmand, J. N., and J. R. Gould, *Technical Reporting,* revised ed., Henry Holt and Company, Inc., New York, 1959.

Van Hagan, C. E., *Report Writer's Handbook,* Prentice-Hall, Inc., Englewood Cliffs, N.J., 1961.

Young, C. E., and E. F. Symonik, *Practical English,* McGraw-Hill Book Company, Inc., New York, 1958.

10

Make Business Travel Pay

. . . in traveling a man must carry knowledge with him, if he would bring home knowledge.
SAMUEL JOHNSON

As an engineer or scientist you probably spend many hours of your working time on the road. For most of today's engineers and scientists the road is one of the nation's airways—like Jet 60 Victor Airway on the Chicago to New York route. Today's technical and scientific personnel probably spend more time aboard commercial airliners than any other profession you can name. Trains, rented autos, and busses also transport thousands of technical personnel every year. But the airplane, commercial and corporate, probably logs more technical passenger miles than all the other transportation facilities combined. For example, engineers of Foster Wheeler Corporation flew more than one million miles during the first half of a recent year.

No matter how you compute the travel mileage of technical personnel, the total is enormous. Converting this mileage into hours also yields a staggering number of man-years. So whether you travel much or little, there are extra work hours available to you. By judicious use of travel time you can accomplish many useful tasks.

Using Travel Time

You can use travel time for a number of purposes: (1) To catch up on job tasks; (2) to further your education; (3) to do personal work; (4) to relax. Since the first three are most important from the standpoint of your career, we'll give them more attention.

Catch up on job tasks. In an airplane, train, or bus you can read or write in comfort. All you need is a hard, smooth surface for writing and you can (*a*) catch up on your correspondence, (*b*) write a report, (*c*) prepare a technical article or paper, or (*d*) even write a professional or scientific book. (Over half this book was written while traveling or waiting to travel). John Markus, well-known electronics engineer, has authored twelve books and shared in the writing of four others—with much of his work done on commuter trains. "A seat on a train's much better than a private office—if you don't have talkative neighbors sitting next to you," Markus says. "And my neighbors are pretty well trained." John Markus coauthored the first dictionary of terms ever published in the electronics field.

What are the keys to catching up on job tasks while traveling? There are two—planning, and preparation.

Planning. Try not to dash off on a business trip without first planning what work you can do while traveling. Take work that doesn't require extensive reference material. Thus, routine correspondence, reports, studies, and the like are easier to do while you travel than many tasks requiring bulky reference material. If you're busy for several days or weeks prior to departure, plan on bringing some routine material with you. This will allow you more time in the office or plant to accomplish those tasks which can't be done away from home.

Prepare. Bring letters, reports, and documents with you on your trip. You can't answer letters if you don't have them before you, unless your memory is better than most. Before leaving, set up a file folder for *Enroute Tasks.* Put all the material you expect to work on in this folder before you leave. For a long trip you might even file material in this folder for several weeks before departure.

Will you need a slide rule? Then bring a 5-inch rule that will easily fit into your pocket. A slide rule will save you many hours of involved arithmetic. What about pencils, pen, paper? If you work with a mechanical pencil, be sure to have extra leads and erasers. Commercial airliners offer many extras, but leads and erasers for mechanical pencils are not yet on the list. Is there a certain type of paper you prefer? Then bring enough sheets or pads to last the entire trip. Having a familiar and favorite paper on hand will enable you to work better in strange places. It will also permit you to organize your work more readily. What about references you may need? If you must bring references, try to reduce them to the minimum size and weight. Don't bring an entire catalog when one page or sheet will do. Lugging heavy references up and down airline ramps will discourage the most ambitious engineer or scientist from working on the road.

Typical useful reference material you can conveniently take on trips includes articles and papers clipped from professional journals, technical and scientific books, reports, and memos. Some engineers and scientists carry their reference material in a separate file from correspondence and other business material. This scheme allows you to organize your material faster.

If you're accustomed to working on a clip board, consider bringing one on your next trip. There are a number of lightweight metal clip boards on the market today.

Long trips. On long trips try these two additional techniques for getting work done: (1) Have work mailed to you; and (2) keep in touch with your secretary or assistant on certain dates at a specific time.

With the first technique you have copies of important letters, memos, and other items mailed to you once a week or more often. Have your secretary enclose a stamped addressed envelope for easy return of the material. Answer important correspondence and other items by jotting your reply directly on the copy of the letter. Then there's no chance of your secretary sending an answer to the wrong person. Use air mail both ways, and you'll save time. By arranging to have your correspondence sent to you while you're away, you avoid that foot-high pile of work when you return. Also, your answers will not be delayed until you return. Be sure to use only copies (carbons, reproductions, etc.) in this system. Then there is no chance of the originals being lost in the mail.

Use the telephone to contact your assistant at regular intervals. Arrange a specific date and time for your call—say Friday, the 20th, at 11 A.M. Then your assistant or secretary will be sure to be available at the phone.

Direct your assistant to prepare a list of items needing your attention. Prepare a similar list of items you want to discuss. When you call your assistant, allow him or her to speak first. Some of the items on his or her list may match those on yours, permitting you to save time.

On long-distance calls speak clearly at all times. Make decisions quickly. Cut details to the bone—where possible; tell your assistant to follow the procedures used in a similar situation. Hold all personal data, weather details, and like information to the last. This way you'll get the important directions across before you or your assistant's mind is distracted by details of California sunshine or New York snowstorms. Before you hang up check your list of items. See that all have been discussed. Also, be sure to ask your assistant, "Is that all? Is there anything else while you're on the line?"

Here are a few hints for more effective business phoning. Keep a pencil and notebook handy while you're talking. If your next few business contacts will be decided at a meeting in the home office, call *after* the meeting, *not* before. This may save an extra call. Close your conversation by giving the date and time of your next call. When doing this, don't overlook any time differences that might exist between your locale and the home office. Lastly, though these hints are particularly useful on long trips, you can also apply them whenever you are away from your desk for any period of time—long or short. Effective use of these mail and phone techniques can keep you in touch with your desk no matter how near or far you may travel.

Other tasks. Enroute business tasks you can perform aren't limited to reading and writing. Other typical tasks include preparing for meetings, reviewing business statistics and projections, analyzing critical situations requiring a decision, etc. Some of these tasks may be best done alone; others are more fruitful when done with an associate from your organization or a related one. Thus, you can discuss business problems with an associate who is traveling with you. Some engineers and scientists actually plan on using enroute time to reach a decision with an important associate who is also making the trip. Many a plan of action, decision, and procedure has been formulated in a plane, train, or bus.

You can also use travel time to become acquainted with associates in your organization. The press of daily activities will often prevent you from becoming familiar with new personnel. On a trip with these people you have time to relax, exchange experiences and knowledge. By carefully choosing your seatmates you can broaden your knowledge of a number of people in your organization.

Make enroute work easy. An important step in preparing any enroute work involves baggage. If you use a large suitcase and travel by air, you'll have to carry your paper work in a separate envelope or dispatch case. For if you leave your paper work in the suitcase, you won't be able to get the papers after the bag is checked in at the baggage weighing desk. For this reason most engineers and scientists today carry two pieces of luggage—a suitcase and an underarm envelope or a dispatch case.

Envelopes are good when you have only a few papers to carry. Stiff envelopes can be used as a writing surface where a table isn't available. If you like to carry an envelope, consider getting an oversized one with several pockets. Then you can put correspondence in one pocket, references in another, reports in a third, etc.

Dispatch cases are probably more popular today than underarm envelopes. Advantages of dispatch cases include neatness, stiff exterior, roominess, and light weight. Some dispatch cases have an alphabetized file inside the lid. One make has an underarm envelope in the lid of the case; another features a small desk blotter inside the lid. The blotter can be taken out and used as a writing surface on top of the lid.

Regardless of your choice—envelope or dispatch case—you'll find either an incentive to doing enroute work. Carrying paper work in a suitcase always leaves you with the excuse that "it's too much trouble to get those papers out now." So instead of working you chat with your seatmate or doze. Why not get yourself a piece of luggage for carrying paper work? You'll be glad you did because your output and efficiency will zoom.

Further your education. All of us in the engineering or scientific professions know how fast things are moving. Professor Derek J. de Solla Price, of Yale University, talking of the acceleration of science says,[1] "For 300 years science has grown exponentially, and the trend is continuing. Manpower can't possibly keep pace; soon there will be far more breakthroughs than we can ever handle." We're getting new knowledge so fast—some estimate that our scientific and technical knowledge doubles about every ten years—that the individual has almost an impossible problem of keeping up.

Professor Thomas Stelson, head of the Civil Engineering Department at the Carnegie Institute of Technology, analyzed the problems of practicing engineers thus:

> Unless a graduate of ten years ago has systematically spent about ten per cent of his time extending his knowledge beyond the level of development achieved in his collegiate training, he will not have value in excess of that of a new graduate. This assumes, too, that he retains all of his previous training, which is probably far from realistic. If decay from neglect or disuse is also ten per cent per year, an engineer is then faced with the task of growing in new knowledge at the rate of about twenty per cent per year to remain of equal value to his employer and society. To increase in value at a significant rate, he should probably devote about one-third of his productive hours to self-education and improvement. Such education is a prodigious task. Its magnitude is seldom realized and all too infrequently attained.

It certainly is a prodigious task. But you can achieve much if you study while you travel. As with business tasks you must plan and prepare if you are to wring maximum benefit from every travel hour.

Plan on reading certain articles, papers, or books on each trip. But remember: You can't read one or more of these items unless you bring them along. So, once again, get a piece of luggage for your paper work *and* studies. Put those important journals or books in the luggage long before you leave. Then you won't forget the material in your rush to catch a plane or train.

When you travel regularly, try to set up a study program for a year, six months, or some other period. You can estimate how many hours will be available to you. With this total in mind you can easily determine how much material you can cover in the period. See Chapter 25 for more details on topics you might consider for enroute study. Lastly, don't overlook the other tools of study—slide rule, pencils, paper, portable drafting set, answer book for problems, etc.

Do personal work. If you travel during working hours—say 9 A.M. to 5 P.M.—you should devote any extra time to your organization first; yourself second. Some engineers and scientists may snicker at this, saying that travel *is* work. Certainly it is—but that is no reason for neglecting your employer's interests. So finish your job responsibilities before turning to personal work.

And what kind of personal work can you do enroute? There are many tasks. Consider these: (*a*) Write technical or scientific articles, papers, or books. (*b*) Do work for an off-hour consulting service. (*c*) Correspond with other scientists or engineers, friends, and relatives. (*d*) Catch up on community duties by reading, writing, or studying important data. (*e*) Sign and insert in envelopes routine letters like requests for funds for your community chest, fire department, alma mater, etc. One well-known engineer headed a $5 million fund drive for his school. Much of his work was done on public transportation facilities. His drive was so successful the school received over $6 million.

As with any other enroute tasks the keys to successful personal work are planning and preparation. Plan your work in advance and prepare the materials you'll need. Then you'll accomplish more in less time.

Where to work. Practice a little, and you can work anywhere enroute—plane, train, bus, or taxi. In the evening your hotel or motel will also provide good working facilities. Almost all hotel and motel rooms have a desk; if the room doesn't, use the lobby. Many hotels will rent or lend you a typewriter, if you need one. Where you must have professionally typed material, use a public stenographer. Most large hotels have one or more public stenographers on their premises.

Don't overlook a portable dictating machine. A number of lightweight models are on the market today. You can use these almost anywhere. Some hotels and motels will even lend you a Dictaphone or Audograph on which you can record your information. Then all you need do is mail the belt or disk to your home office for transcribing.

Talks with a large number of engineers and scientists show that most of them prefer to work with pencil or pen and paper. Probably the reason for this is that most of these men believe they can be more exact when writing in longhand. So if you prefer longhand over the typewriter or dictating machine, don't be alarmed. You'll probably accomplish more than with unfamiliar methods.

When seeking places to work don't overlook airport terminals, train and bus stations. A quiet spot in any of these can be an ideal place to work. And during those long waits at airline gates have some work ready. You'll keep your place in line, the time will pass more quickly—and you'll get some work done.

How to work while traveling. Here are some useful hints for making work easier while traveling:

1. Pick a comfortable seat with adequate light and ventilation. Avoid drafty areas near doors and other openings.
2. Steer clear of groups of people traveling together, particularly children. They'll probably annoy you during the trip, preventing concentration.
3. Try not to sit next to talkative amateur travelers—they'll resent your lack of attention to them. Study the passengers you see on your next few trips. You'll soon be able to spot the talkative types.
4. If you plan to work, *don't* buy that detective story, favorite magazine, or other periodical before you leave. For if you do, the work is almost certain to be neglected. Wait until you finish the work—then reward yourself by buying a favorite author's story or a magazine you enjoy.
5. Stay away from magazine racks in planes and trains until your work is finished.
6. Try to avoid accepting a drink until your work is finished. Alcohol has strange effects at 30,000 feet.
7. Nap after working—not before. The nap will refresh you more after your work is finished.
8. Use every travel opportunity to work. Dr. William A. Geyger, of the Magnetics Division of the U.S. Naval Ordnance Laboratory at Silver Spring, Maryland, uses the time he spends as a passenger in a commuter auto to study the latest patents in his speciality—magnetic amplifiers. Extensive study in this field led to publication of a fine book on magnetic amplifiers which is now in its second edition.
9. Take advantage of subway and bus travel. You may spend several hours a week on subway trains or busses, if you live in a metropolitan area. Learning how to work in these conveyances can increase your output considerably.
10. Carry a dated diary or appointment book with you when you travel. By noting important work and appointments in the diary you can keep account of what is to be done and what you accomplished.

Use travel to relax. Once you've finished your job and personal tasks, relax. Your travel time will be much more rewarding if you arrive at your destination relaxed and ready for work. So learn to stretch out and nap in any conveyance—plane, train, bus. Here are a few hints for more relaxation during your travels:

1. Remove eyeglasses and loosen your shoes before taking a nap. This way you'll rest better.

2. In planes and busses turn off or direct the ventilating air away from your head before you take a nap. Otherwise you may catch a cold while asleep.

3. On long jet flights ask the stewardess for a pair of the knitted slippers provided free to first-class passengers by United, American, TWA, and other airlines. These slippers are cozy and seem to induce sleep.

4. If you find poker, gin rummy, or bridge more relaxing than a nap, look around for a few players. You can almost always find a few who will gladly join you. As a last resort, you can play solitaire if you prepare and bring a deck of cards with you.

5. When your usual interests pall—as they sometimes will on long trips—turn to the travel folders and maps. These interesting items will often give you useful suggestions for short visits you can make to historical and other exhibits at your destination, *after* working hours. Take the advice of one engineer who's been in and out of every important and unimportant airfield in the United States: *Business travel can be productive and interesting—if you plan and prepare.* As Francis Bacon said, "Travel, in the younger sort, is a part of education; in the elder, a part of experience."

Checklist of Travel Supplies

1. Business cards
2. Transportation tickets
3. Confirmation of hotel reservations
4. List (names and addresses) of people to see
5. Copies of correspondence related to trip
6. Business reading matter (letters, memos, reports, etc.)
7. Paper, pencils, pens
8. Slide rule
9. Expense account forms
10. Credit cards

REFERENCES

[1] D. J. de Solla Price, "The Acceleration of Science," *Product Engineering*, March 6, 1961.

11

Use Spare-time Activities for Growth

> No man is really happy or safe without a hobby, and it makes precious little difference what the outside interest may be—botany, beetles or butterflies, roses, tulips or irises; fishing, mountaineering or antiquities—anything will do so long as he straddles a hobby and rides it hard.
>
> SIR WILLIAM OSLER

Your hobbies can help you grow as an individual—in mind as well as body. Too many engineers and scientists stop growing as soon as they leave school. Their minds, except for the specialty in which they work, become sluggish. These engineers and scientists overlook Samuel Johnson's famous remark—"Curiosity is one of the permanent and certain characteristics of a vigorous mind." Too routine a life can stifle curiosity and stunt your personal and professional development.

Don't Become a Slave to Routine

Being logical and precise, as most engineers and scientists tend to be, you may be easily victimized by routine. Once you acquire a home and begin to raise a family your life takes on a regular pattern. You go and come by the clock, with little variation. As the years pass, much of your zest for life seems to disappear. You accept your family and home as comfortable parts of your routine. Your career moves ahead slowly—far more slowly than you ever thought it would.

Then a certain birthday gets you thinking. It may be your thirtieth, fortieth, or later. You take stock. With a startling suddenness you realize that you've allowed your life to become a treadmill. You've centered yourself between your job and your family. Your family, and perhaps your job, grow. But you remain static between the ears and as a personality.

Now certainly you must look after your family and your job. Your first responsibility is to them—and a certain amount of routine helps make life easier. But if you allow this routine to spread to your mind so your sense of growth is deadened, you soon begin to miss some of the best parts of life.

Check on yourself now. Are you a slave to routine, as so many other engineers and scientists are? Ask yourself a few questions:

1. In what hobby are you an active participant?
2. Do you have any major interest outside your job and family that fills your daily thoughts?
3. Do you spend time, money, and energy on this interest?
4. Have you ever taken an active part in your local community activities (town government, school board, PTA, Boy Scouts, etc.)?
5. Are you active in any of the arts—(painting, music, theatre, sculpture, etc.)?
6. When did you last take a course in a nontechnical subject (history, management, human relations, a foreign language, etc.)?
7. Do you ever travel for pleasure?
8. Have you read a *good* novel, poem, or play since your last course in English or American Lit? [1]
9. Have you truly tried to become interested in people and what motivates them?
10. Are you conscious of your appearance, clothes, health, speaking ability? Do you regularly make a sincere effort to improve these and similar aspects of your life?

If you can't answer an objective yes to at least five of these questions, you may be a victim of routine. Unless you begin to do something about your personal habits you will continue to live a stunted mental and physical life. Though you may not realize it, limiting your spare-time activities can retard your professional achievements. The reason this is so is because without spare-time interests your personality becomes narrower, and within a few years your outlook is confined to a very small part of the world. You can avoid this narrow outlook. Let's see how.

Four Aids to Growth

Every engineer and scientist has many opportunities to achieve more in his profession. At the same time he can grow in capability, maturity, and experience, if he leads a balanced life. To counteract the narrow specialization of your daily activities you need the following four aids to growth. These are: people, places, personality, and preoccupation.

People. Unless you're the unusual engineer or scientist, you have problems with people. You're shy, embarrassed, and tongue-tied with almost all people except fellow engineers and scientists. You're aware of this deficiency but cannot seem to do much to cure it.

Why do you have this problem? Mainly because of your technical training, which has limited your world to a relatively narrow area. You studied technology so intently that many of your other interests were neglected. Today you have little you can talk about to businessmen, medical doctors, accountants, laborers, civil-service employees, etc. Why? Because they don't speak your language of engineering or science and you can't speak the language of their specialty. But what's worse yet, you can't or find it difficult to speak to these people as human beings.

How can you be friendlier, more relaxed with people? Try these three tips. (1) Have *something* to talk about—your hobby, an interest in sports, trips you've taken, etc. (2) If you don't have an interest, other than engineering or science that you can talk about, get one. Were you interested in baseball, skiing, track, or basketball as a boy? Renew this interest. Get some good books on the subject. Learn what has happened since you dropped the sport. The same goes for stamps, tropical fish, model airplanes. Buy a copy of the magazine serving your interest. See how much fun it is to read. (3) Seek out others with your interest—preferably not among your engineering and scientific friends. Compare views about your interests. Steer clear of shoptalk about your job—concentrate on the hobby or interest.

Listen intently to the next few nonbusiness conversations you hear. Note how much of each conversation revolves around hobbies and other nonjob interests. You can easily take part in such conversations when you have interests that are broad enough to encompass more than your job. Meeting people on these terms helps you be friendlier, more relaxed. This attitude will carry over to your professional work and will help you there. What's more, you'll learn more about the world, and help others learn more about it when you exchange information about hobbies and other interests.

Places. As an engineer or scientist you probably travel to conventions, meetings, and to other organizations with which your employer works. So you get to see many places other than your local work area. But what about the places closest to you—those in your community? Do you visit your local PTA meetings, Community Chest, fund-raising drives, etc? In these places, so close to home, you can make a real contribution by working for a cause in which you believe. And working for this cause will bring you into contact with people in your age group but different professions and occupations. Meeting and working with these people will take you out of the daily routine of your job and show you how to get along well with persons having diversified backgrounds.

Don't overlook other places offering new and broadening experiences—the theatre, art museums, town meetings, even the zoo. Each of these, and many others, offer unique chances for you to participate in worthwhile activities. So choose those places you think would be most interesting to you and get to work. Remember what A. Edward Newton said, "... get a hobby; preferably get two, one for indoors and one for out; get a pair of hobby-horses that can safely be ridden in opposite directions." Try his advice when choosing your places for spare-time activities.

Personality. Almost everyone has personality problems of some kind. Some people are bashful; others are domineering. As you've matured, certain of your own personality problems have come into focus. You might realize, for example, that you can't delegate work to your subordinates. You want to keep things under control. So you try to do everything yourself, even the routine work that could be done by an able assistant. "Someday," you say, "I'll begin delegating work." But that someday never arrives.

Richard P. Calhoon, while personnel director at Kendall Mills, wrote the following:

> A good part of your success centers around your personality—how you affect others. . . . Personality is more important than technical skill. Technical skill is a necessary basis for certain types of work, but many more failures are due to personality failings than to weaknesses in technical skill. . . . You are never quite certain how your personality functions because results are dependent on others. You have a pretty good idea, however, from objective facts. In a group of three to five, does the one speaking address himself to you as much as to the rest? Do people confide in you? How generally are your suggestions adopted by a group? Are you ever elected to office? How often are you asked to serve on committees? If you answer none of these questions with a yes, you should probably be considering what will improve your personality.[2]

Don't be weak-willed—develop a strong personality. Most leaders in engineering and science have a strong belief in the importance of their work. They pursue their goals vigorously, while maintaining an understanding and fair attitude toward their associates and subordinates. The days of the eccentric scientist working in his makeshift laboratory are gone. Today, more than ever before, industry and government demand that their engineers and scientists be individuals with well-balanced personalities and a sense of dedication for their work.

Check up on yourself.[2] Are you (1) overbearing, (2) superficial, (3) wishy-washy, or (4) zero (the blah type of person who makes few or no positive or negative impressions on people)? Think for a while. You can *sense* your major personality lacks. Do something about them.

The best cure for personality problems is people. For the more you work with people the easier it is to see what you may lack. And by taking positive action to overcome these problems you can advance to greater accomplishments in your professional and spare-time activities. So begin now to search for your personality lacks. Having found them, use your hobbies to help you overcome the problems.

Preoccupation. You've probably laughed at the sports-car buff, golf enthusiast, or fishing "nut." Why do people laugh, in a gentle way, at these hobbyists? Because the hobbyist seems to be a man in love—his full attention is given to his hobby whenever he has a free moment. He reads about his hobby, practices, visits stores featuring supplies, talks endlessly to fellow hobbyists. He resembles a man in a trance.

Is such an interest good? Or does it harm a man's professional contributions? Every study ever made of man and his hobbies indicates that intense preoccupation with an external interest is an excellent tonic. A hobby renews a man's energy, helping him do a better job when he returns to his daily activities. So never carry a "hobby guilt." In fact the more time and energy you put into a hobby, the greater your return in the renewal of your desire to accomplish more in the world of engineering or science. Any hobby will keep you alert, interested, and interesting.

The Future Is Yours

You have a whole new world open to you when you decide to use spare-time activities for personal growth. Recognize today that you can, if you wish, broaden your interests and accomplishments. Doing so will deepen your appreciation for life, for your profession, for your family, and for the importance of human effort. As an engineer or scientist you are in an excellent position to make lasting contributions to your fields of interest. Begin now to expand and enlarge your spare-time activities to build a greater future for yourself and your family.

An Associate Editor of *Mechanical Engineering* once said, "The engineer should be a man of the world, knowledgeable in philosophy and in the political facts of life. He falls painfully short of this goal.... He tends to believe that the world's problems can be solved by more efficient machinery—a convenient belief, but one that may cost him the greatness he ought to achieve." [3]

REFERENCES

[1] The Engineers Council for Professional Development, *Selected Reading for Young Engineers,* ECPD, United Engineering Center, New York 17, N.Y.

[2] Richard P. Calhoon, *Moving Ahead on Your Job,* McGraw-Hill Book Company, Inc., New York, 1946.

[3] Herbert B. Devries, "Have You Read . . . ?" *Mechanical Engineering,* February, 1962.

HELPFUL READING

Dickson, F. A., *2000 Articles You Can Write and Sell,* Perennial Press, New York, 1955.

Kingery, R. E., *How-to-do-it Books,* 2d ed., R. R. Bowker Company, New York, 1954.

Lovell, E. C., and R. M. Hall, *Index to Handicrafts, Model-making, and Workshop Projects,* F. W. Faxon Co., Inc., Boston, Mass., 1950.

Smith, F. S., *Know-How Books,* R. R. Bowker Company, New York, 1957.

12

Should You Change Jobs?

> The circumstances of others seem good to us, while ours seem good to others.
>
> SYRUS

Today engineers and scientists are probably the most mobile professional group in industry. A survey by *Space/Aeronautics* magazine [1] showed that in one group of engineers 43 per cent were "wondering whether they should change jobs." Further, "Engineers don't seem to lose their mobility as they get older. Almost half of the men in their forties (in this survey), as compared with exactly half of the men in their twenties, are thinking about changing jobs. Another survey shows that scientists change jobs on the average of once every 3.3 years. This would give a turnover rate of about 30 per cent per year for the average scientific organization covered by the survey.[2]

Why Mobility?

What causes engineers and scientists to job hop so much? Is it money? The *Space/Aeronautics* survey shows ". . . the correlation between an enginer's restlessness and the size of his paycheck seems to be only moderately high. Over half the engineers in our survey . . . are satisfied with their present salaries and . . . 38 per cent are thinking of changing jobs anyway." Interestingly, "On the average, they'd move (change jobs) for a 20 per cent raise."

Other factors appear as important as, or almost as important as, earnings. The same survey found that, "Almost as many complaints are voiced about the lack of authority to go with the engineer's responsibilities, about lacking opportunities for technical creativity, and about harassment by petty rules and regulations."

Another survey, by Opinion Research Corporation, verifies the *Space/Aeronautics* survey. Thus, O.R.C. found the following:

> In a survey of 622 engineers and scientists, 72 per cent thought management misused their talent, and 80 per cent said they were underpaid compared to other groups . . . most thought their goal in the company was pursuit of knowledge for its own sake and wanted less pressure from management, freedom to select their own work, flexible work schedule . . . managers interviewed showed good understanding of what engineers and scientists wanted, but called it a type of freedom that belongs to academic institutions, not to competitive industry.[3]

There you have it—many engineers and scientists believe they are misunderstood. So they job hop, looking for the optimum spot—one with adequate income, job freedom, intellectual atmosphere, good living conditions, superb schools for their kids, etc. They are striving to escape what they call [3] "second-class citizenship in the professions," "a second-class profession," "a subprofessional group," and being "treated like 'peons' or 'laborers.'"

Does job hopping pay off? Should you change jobs? There isn't a pat answer to these questions—much depends on your professional goals, income needs and desires, climate preferences, etc. Some engineers are able to pyramid one raise on top of another by job hopping from one firm to another. This technique is popular in the electronics and aerospace industries. A serious disadvantage of pyramiding is the danger of being classed as a "floater" by future employers. Most firms shun the floater because his past history shows that he won't stay long when he's hired.

Why Change Jobs?

Examine your motives for making a change before you even hint of resigning. Use the following checklist to rate your degree of dissatisfaction with your present position. Be realistic in your answers —don't check off "salary too low" just because you dream of earning $1 million per year. Check off this factor only when you sincerely believe you are underpaid or your normal living expenses greatly exceed your income.

	YES	NO	
1.	_____	_____	Is your salary too low?
2.	_____	_____	Have you received regular raises?
3.	_____	_____	Can you ever expect to earn much more than at present?
4.	_____	_____	Has your supervisor promised you future promotions?
5.	_____	_____	Are you reasonably happy in your current position?
6.	_____	_____	Do any top men in your organization have the same education as you?
7.	_____	_____	Can you reasonably expect to ever be considered for a high-level job?
8.	_____	_____	Are your working conditions tolerable?
9.	_____	_____	Is transportation to and from work difficult, unpleasant?
10.	_____	_____	Do you like the general weather in the area?
11.	_____	_____	Is the school system in the area satisfactory?
12.	_____	_____	Are there suitable cultural activities in the area?
13.	_____	_____	Do you have opportunities for professional education in the area?
14.	_____	_____	Is the organization well-managed?
15.	_____	_____	Does the organization's main activity have growth possibilities?
16.	_____	_____	Is the organization consistently profitable?
17.	_____	_____	Are layoffs of people in your category common?
18.	_____	_____	Do you feel reasonably secure in your present position?
19.	_____	_____	Are your fringe benefits (hospitalization, pension, etc.) sufficient?
20.	_____	_____	Is your organization contemplating major changes (merger, sale, etc.)?
21.	_____	_____	Do you get along reasonably well with your associates at work?

	YES	NO	
22.	______	______	Would you and your family really be happy in another location?
23.	______	______	Does your present position give you a chance to be creative?
24.	______	______	Are people in your category respected in your organization?
25.	______	______	Do you have any serious complaints about the organization?
26.	______	______	Are you able to make full use of your talents in your job?
27.	______	______	Can you expect to move ahead nearly as fast as you wish?
28.	______	______	Is your work interesting and challenging?
29.	______	______	Do you believe you are on the way towards achieving your major goals?
30.	______	______	Do you have a major goal that you actively pursue?

Score yourself in the above checklist as objectively as possible. Count the number of negative answers. If you have fifteen or more, you're not very happy in your present spot. But don't quit today—look around to see if you can make a change *within your present organization.* Remember: There is no perfect job so look carefully at your advantages before making a change to an unknown organization and new set of working conditions.

If you find twenty to twenty-five *No* answers you should seriously consider making a change. For having a job with that many imperfections can do nothing but make you irritable, unhappy, and morose. Your professional achievements will be minor, or nonexistent.

Plan Your Change

T. K. Kelly, while a Chemical Engineer with American Machine & Foundry Company, gave some useful hints on changing jobs.[4] His thoughts are directed at all technical personnel who have the ability to advance rapidly but who may lack the decisiveness needed to grasp an excellent opportunity.

Looking into job opportunities outside your organization may, if nothing else, shed new and encouraging light on your present position. While looking around may prove that your present salary is below average for your particular professional training and technical or scientific skill, it may also show that your employer is not treating you as badly as you think he is. The grass may look greener on the other side of the fence—but colors are deceiving.

When roots are firmly established in a particular local area, when you have built up a circle of friends, own—or are mortgaged to—your home, and have your children in a good school, it's extremely difficult to uproot the pattern of life that you've taken years to establish. Unless the future in your new job horizon has very positive advantages for you and your family, your resistance to change probably will be very strong. However, if it is to your advantage to make a change of employer, then the earlier you recognize the true situation, the less adjustment will be necessary when you relocate.

You surely need a planned, scientific approach to changing your job. At best the process of selecting a new employer and of actually making a job change is a difficult and at times a discouraging task. One chemical engineer, after he had voluntarily relocated, said, "Finding a new position was one of the most gruelling experiences of my life. The interviews, agencies, consultants, résumés, and all the complicated forms made my head spin. I got a better job and am happier for it. But I've been through the mill. And believe me—it's no fun."

Since the first step in the scientific method is to collect whatever information already exists in the field of study, the information in this chapter will help you take step number one. Much of the information presented will also simplify the steps in the relocation procedure. This information may also help you eliminate some unnecessary steps.

What to expect. Here are some general rules of the road that usually hold for today's journey into the job market: (1) Figure on at least three to six months between the initial inquiry and actual placement. (2) Never wait for an answer from one application; apply for as many positions as you think you can fill. (3) While looking for a new position, never let down in carrying out assignments to the best of your ability for your present employer—you may decide to stay with him. (4) Beware of vague advertisements that omit company name and location. Some blind advertisements are run as a survey for the availability of persons with specific training in preparation for future or indefinite expansion of a particular company or industry. Note also that this preliminary screening will protect you against applying for a job with your present employer. (5) Keep a file folder of all correspondence, applications, and replies to inquiries.

CONFIDENTIAL

RESUME

OF

JOHN R. CAPABLE

Residence:
21 Amosland Road
Bridgeville Manor, N.Y.
BR 5-1234-Code 516

Born U.S.A. 1925
BME - Lehigh
MME - Geneva

OBJECTIVE: Engineering Position at the Executive Level

FIELD: Electronics and related industries

QUALIFICATIONS: Skilled and experienced in all phases of mechanical engineering, production and design.

Effective in building sales and profits for the company. Proficient in the design and manufacture of electronic and electro-mechanical equipment and instruments.

A competent manager in the supervision of the work of others. Gain cooperation ... meet work schedules ... operate economically.

Thoroughly conversant with the philosophies and techniques of domestic and foreign business men - conform to their preferences and requirements.

Particularly effective in the development of Preventive Maintenance procedures and the standardization of the uses of Test Equipment by technicians at scattered sites. This reduces complaints and keeps costly scrap and rework to a minimum.

EXPERIENCE, ACCOMPLISHMENTS and BACKGROUND:

1955 to Date: With a large instrument manufacturer in the New York metropolitan area, whose name will be furnished at an interview.

Position: Superintendent Engineering.
Duties: Supervision and administration of engineering and design staff, the laboratories, the model and production shop and assembly floor. Submit bids to prospects and customers - United States Armed Forces, Atomic Energy Commission and other U. S. Bureaus.
Products: Precision components, switches and controls for the guidance of missiles and the operation of space vehicles.

FIG. 12.1. Typical resume. (*Courtesy of Robert B. Buck and the Executive Counseling Service, Inc.*)

Prepare a résumé. Your first step in hunting a new job is preparation of a concise and attractive résumé (Figure 12-1). Be sure the résumé is reproducible. If you need help in preparing your résumé, check the yellow pages of your phone book. Most large cities have several firms specializing in the preparation and reproduction of résumés. Their charge is reasonable and the results effective because

CONFIDENTIAL

Resume of John R. Capable — Page 2

EXPERIENCE, ACCOMPLISHMENTS and BACKGROUND (continued)

1955 to Date (continued)	Accomplishments: Reduced costs resulting from a redesigning of certain products enabled us to effect improvements and establish more competitive prices. Result: we are now in a dominant position in our market.
1953 - 1955	TODAY AND TOMORROW CLOCK CO., Earleyville, N. Y. Manufacturers of Timing Mechanisms for every purpose. Position: Project and Design Engineer. Accomplishments: Collaborated with Electronics Department in the design and production of new timing devices for the U.S. Navy, the Merchant Marine, and commercial organizations. Result: Today, the company is enjoying substantial profits from patents of devices which I designed.
EDUCATION:	BME degree, Lehigh University, 1949. MME degree, Institute of Technology, Geneva, Switzerland, 1952. While completing education, and in accordance with University requirements, acquired a thorough knowledge of modern machine shop practice and machine tools in various shops.
MILITARY:	Lieutenant, USNR All military obligations completed.
LANGUAGES:	Speak, read and write: German, French, Spanish.
PERSONAL:	Married. Height 6' - weight 180 Willing to relocate. Am now bonded by my company. Secret clearance.
REFERENCES:	Upon request.

you obtain a concisely worded résumé of professional quality.

Send a covering letter with each résumé when answering an advertisement. Keep the letter short—not more than one page—and do not repeat information given in the résumé. Here are some useful hints to guide you in preparing your résumé.

Divide it into two parts. Give personal details first on a separate sheet. These details should include your name, address, telephone number, age, education, degrees, military service, present draft status, clearance classification. Also list your present salary and three or more business references. Be sure to obtain approval from your references before listing them. Some people dislike giving a recommendation unless asked in advance.

List the telephone number of each reference in the résumé. Telephone checks of business references are popular today. The telephone check is quick, inexpensive, and accurate. Answers given over the phone are often more frank than those committed to writing. Also, the person doing the checking can ask precisely the questions applicable to the job being filled.

If possible, use former supervisors as references. These men will be familiar with your working habits and ability to cooperate as part of a team. Today many employers stress a man's teamwork ability because so many important tasks are handled by teams of engineers and scientists.

In the second part of your résumé list work experience, starting with your current position. For each position include your job title or titles, highlights of your accomplishments, and levels of responsibility. Be brief in describing each position, but *don't undersell yourself.*

What happens next? The first response from an organization interested in you may be an application blank to be filled out and returned. If the application is satisfactory, you will be invited to visit the organization.

You may or may not be asked to attach a photograph to your application blank. This is where you may run into both a legal and a diplomatic problem. First, you certainly will not be hired without a personal interview, so the request for a photograph can only be used as a negative selection device. The only use for the photograph is to eliminate those candidates whose appearance is unsatisfactory.

The legal problem arises in those states having fair employment practices laws or state commissions against discrimination. For example, an employer in New York State is acting illegally if he asks you to submit a photograph as part of an application for employment. If you submit one, you might be aiding him in an illegal act.

To avoid these problems, many engineers and scientists ignore a request to submit a photo with their application. And prospective employers seem just as interested in you, photo or not, if you have the skills they want.

How to be interviewed. Count on spending one full day for the job interview with each prospective employer. Your transportation to and from the place of interview will usually be paid by the employer.

The interview is a job of selling yourself. Be sure you have a well-groomed appearance, even though you've had a long trip to the interview site. For best results in the interview, keep talking about your background. Always speak or look directly at the person conducting the interview.

Remember that your interviewer wants to ask the questions and prefers this to answering your question. But he will expect you to ask for details about salary, overtime pay, vacation, benefits, relocation expenses, amount of travel you're expected to do, and organization growth potentials. Most large firms have booklets prepared to give you the answers to many of these questions.

Prepare in advance to answer this key question: Why do you want to leave your present employer? Answers can include: (*a*) lack of growth opportunities, (*b*) interest in relocating in a particular geographical area, (*c*) higher salary, (*d*) greater freedom to use your skills, etc.

From you the prospective employer likes to hear this question: Where can I expect to be in five years? Asking this question indicates you are interested in growing in the organization and in assuming greater responsibilities. The employer's answer to this question will be a good lead on the possibilities of advancement for someone with your qualifications.

Make a prompt decision. Every organization will expect an answer to their offer—if they make you one—within a week to ten days after your interview. Do not accept an offer of employment without a personal inspection of the organization's location and tours of its plants, laboratory, etc. Sometimes one look is worth a thousand words and pictures.

If you decide to accept the offer, take extreme care to maintain a cordial attitude toward your present employer. Do not voice opinions that will create a feeling of animosity between your supervisor and yourself. (You never know—within a few years you may be back with your present organization in a much better position.) There's a right way and a wrong way to quit a job. Always (1) give sufficient notice for a replacement to be selected—two weeks is the minimum notice your employer expects; if possible, give him at least a month's notice. (2) Cooperate when your present employer asks for the true reason for your leaving—tell him why, but use polite terms. Remember that in giving this information you may be helping your employer create better working conditions for other engineers and scientists. By being cooperative you help the entire profession advance.

Where to look. Here's a list of twelve places where you might find a new position. You've probably heard of almost all of them at one time or another but having them in list form will be convenient for you. So be sure to check: (1) technical journal ads giving the company name and location, (2) newspaper ads giving company name and location (*New York Times, Los Angeles Times, Houston Post,* etc. are good sources of employment ads), (3) professional-society employment agencies, (4) career centers held in conjunction with society meetings, (5) commercial employment agencies specializing in engineering and scientific personnel, (6) friends and associates who know your abilities and interests, (7) government (Federal and state) agencies needing personnel, (8) research institutes like Stanford Research Institute, Midwest Research Institute, etc., (9) educational institutions like colleges, universities, technical institutes, high schools, and trade schools, (10) staffs of professional societies (many graduate engineers and scientists work on the staffs of professional societies), (11) engineering and scientific trade associations, (12) public relations firms specializing in engineering and science.

Some general hints. Here are twelve hints that will make your job hunting more effective. Try to use as many of these hints as possible. (1) Prepare a good résumé—or pay to have a good one prepared—and keep it in the mail. (2) Send out at least 100 résumés to the same number of organizations. If you think there's the slightest chance of getting results from an additional 100, send them out. Paper and postage are cheap compared with the results one résumé in the right spot can bring. (3) Be aggressive in seeking your new position. If it's 4:30 in the afternoon and you feel like quitting job seeking for the day, make one more call. Some of the best jobs have been found between 4 and 6 P.M. (4) Stretch your imagination—use the checklist in the previous paragraph for job-source ideas. Add other sources to it. (5) Be creative in your job hunt. The creative applicant often has a better chance than the man who's bound by routine. (6) Seek the new industries when job seeking. Go to the companies that are pioneering—they almost always need good engineers and scientists. (7) After each interview write immediately, thanking the person you saw and telling him why you'd like to work for his company. A letter like this will often clinch the job for you. (8) Keep pushing—the more companies you see the greater your chances of obtaining the right job at the best salary. (9) Start on the right foot when you get your new job. (See the earlier chapters in this book.) (10) Never forget that job hunting is a gruelling task. Approach it properly, and you have much to gain, nothing to lose. (11) Use your ingenuity—include reprints of your articles, advertising circulars for your books, society-meeting programs in which you have taken part, with your résumé. Items like this often make the difference between yes and no. (12) Place an advertisement in one of the technical journals, preferably a weekly,[5] serving your field. It will take from two to six weeks before the ad appears in the journal. Here's a typical ad:

> Chemical engineer, age 26, married. Four years experience process development, plant engineering, plant startup and production including supervision. Seeks responsible position in production supervision or project engineering. Will relocate.

This particular ad drew ten replies, four interviews, and three offers in the depths of a serious recession. This is a much higher average than any other single source of contacts.

Overseas Employment

At one time or another during his career almost every engineer and scientist idly wonders about overseas employment. The lure of a

Caribbean or Mediterranean sunset can be a strong element in a man's decision. Extended contact with a distant culture can offer educational advantages and give a man a cosmopolitan background. Are you a candidate for overseas employment? Read on and see.

Foster Wheeler Corporation is typical of a large number of American concerns that ship "expatriate" employees from continent to continent.[6] But unlike many companies that use complex techniques to test whether personnel fit overseas tasks, F-W is more interested in whether its men know any engineering than in how many languages they speak. The company briefs its engineers on countries—without subjecting them to a course on North African Anthropology, or making them learn Swahili in a few hours. More than 10 per cent of the F-W payroll is composed of overseas employees. These men form a very special group in the company.

Typical overseas engineer is Carl R. Wenz. Married, his permanent home is California, but after working for F-W in Indiana for two years, he was sent to Iran to supervise the erection of a refinery, then to Trinidad, then to Columbia for three years, then to Ontario, then to Bataan in the Philippines.

The last was a two-year job involving the clearing of 600 acres of jungle, the management of some 4,500 workers, and the building of a $45-million refinery for Standard-Vacuum Refinery Company (Philippines). It also involved a host of special problems not normally arising in the company's Fifth Avenue office.

From the Philippines Wenz moved on to Turkey where he coordinated the work of twenty other U.S. expatriates, plus sixty supervisors from the United Kingdom, and more than 3,000 Turkish employees. On top of all the difficulties posed by any large construction job, Wenz and his associates had to remember not to refer to Turks as "natives." The Americans had to ship in their own linens, cooking, and household equipment—even their favorite razor blades. They learned to like Turkish coffee and cigarettes.

Recreation in these distant spots can be a major difficulty, especially for the 50 per cent or more of the men who bring their families. In Mersin, Turkey, an unfinished hotel nearby was leased and converted to a club for "ex-pats," with a swimming pool, tennis courts, pool room, movies once a week, and dancing Saturday night.

Typical overseas conditions. Here are some general facts about overseas employment that will help you decide whether you'd like to "go foreign."

(1) Companies like F-W brief each man about overseas conditions before sending him out. These briefings range from the presence of the common cold to the frequency of earthquakes in the area. (2) You may be required to have all dental work completed before leaving the United States because of lack of facilities overseas. (3) Simple items, like Band-Aids, may be unobtainable, so you must bring a good supply. (4) In some climates simple cuts take longer to heal than in the U.S. (5) Storms can frequently disrupt telephone service. (6) In some climates washable suits are better than nylon suits, the latter being too hot. (7) You may need a two-year supply of shoes and shoe trees. (8) In some areas centipedes and scorpions are occasionally found in the house, snakes in the yard. (9) Records and the phonograph provide most of the entertainment for music lovers. (10) Some countries have no appropriate schools beyond the ninth grade. (11) In others, almost everyone develops heat rash from time to time. (12) With Foster Wheeler some men on overseas assignments are on a regular salary, but most accept individual contracts for each job. (13) Where an assignment lasts a year or more, the chances are good that the man will bring his wife and family with him. Most companies believe this is a wise decision.

Make a careful study of yourself, your family, and your needs and wants before accepting an overseas assignment. Be certain your wife fully understands the conditions she and the children will meet overseas. Compare the advantages of being overseas, out of touch with the mainstream of your firm, with the advantages of working in the main office or a domestic branch. If you still hunger for a stint overseas, then go ahead and try it—wife and kids willing. And don't worry about your age—most firms doing overseas work prefer seasoned engineers and scientists who (*a*) have proven competence, (*b*) have good personal reputations, (*c*) can carry the heavy responsibilities of leadership of foreign nationals, and (*d*) can live and work far beyond the reach of most of the comforts and conveniences of civilization. Financial remuneration in overseas assignments generally exceeds that for the same position in the United States.

Should You Be an Individualist?

In seeking a new job at home or abroad which trait should a man stress—his individuality or his conformity? For the last few years a man would have been safer if he stressed his ability and willingness to conform. Recently, however, there has been a swing away from the organization man in many corporations. For example, the *Wall Street Journal* reports: [7]

> A few years ago the personnel director of Chance Vought Aircraft, Inc., suddenly withdrew a lucrative job offer he had made to an experienced 40-year-old executive employed by a competitor. Though the executive appeared highly qualified and was a leader in his field, the results of a personality test supposedly had revealed him to be "emotionally unstable and insecure."
>
> Today there's not a personality test to be found in Chance Vought files; in fact, the results of all such tests given in the past have been deliberately burned. Moreover, the once rejected executive has since been hired and has risen to the ranks of top management at Chance Vought, now a subsidiary of Ling-Temco-Vought, Inc.

This report points out further:

> That transformation in Chance Vought's thinking about what it takes to make a good executive is being duplicated at many other companies around the nation these days . . . there are signs an increasing number of companies are becoming disenchanted with the conformity-minded organization man and instead are placing new stress on individuality and originality in executives. While such men may on occasion ruffle feathers in management ranks, the companies are concluding their contributions generally more than compensate.

The same article cites a number of executives and firms favoring the individualist:

> Gifford K. Johnson, President, Ling-Temco-Vought, Inc.: "There's plenty of room in our company for the bold, brash individual who's willing to be set apart from the herd."
>
> John L. Hardy, a New York executive recruiter: "More than ever before industry is seeking men of originality with the courage to approach problems from an unorthodox standpoint."
>
> Remington-Rand Corporation also "is encouraging more individual initiative . . . the company asked . . . men . . . to take a critical look at their duties. There were a number of surprising results. Most startling of all, four men came up with the original suggestion that their jobs be eliminated."
>
> One company which has taken extreme steps to promote individ-

ual initiative is Texas Instruments, Inc. . . . Texas Instruments ignores organization charts because "people, not organizations, get things done," says Dr. Scott Myers, director of personnel relations . . . the seeming indifference to normal corporate etiquette and protocol is part of a carefully thought-out plan to stimulate imagination and creativity.

Thus, the opportunities to express your individuality are increasing. The wise engineer or scientist, however, does not overextend his individuality. He keeps himself in check until he's reasonably sure of where he stands.

What If You're Fired?

What should you do if you've been let go, laid off, permitted to resign, furloughed or—bluntly and plainly—fired? P. J. Brennan, while assistant editor of *Chemical Engineering,* gave some excellent suggestions.[5] These were:

1. Go home and tell your wife. Her understanding and sympathy will buoy you up at a time when you're in need of some help.
2. Check with your local office of unemployment insurance to see if you are eligible for benefits. Don't let false pride prevent you from accepting these benefits. If you are eligible for them, take them. Looking for a job is a full-time, expensive business for which you'll need every nickel you can lay your hands on.
3. Use the federal or state job agencies. Recognize, from the outset, that it will take at least a month from the time you begin your job-seeking campaign until you get your first acceptable offer.
4. Prepare your résumé. Use the suggestions given earlier in this chapter.
5. Use the 12 suggestions you were given earlier in this chapter to make job contacts.
6. Keep looking, day and night. Don't give up—you'll get a good job if you actively search.

Lastly, don't be caught off-base when you're fired. Read the next paragraph and do exactly what it says *immediately* after you finish reading the paragraph. Taking the action suggested will halve your worries about being fired.

Prepare a comprehensive list of where you'd look for a new job if you were fired today. List the names, telephone numbers, and ad-

dresses of the people you'd contact. Fill out a file card for each job opportunity. List the name of the position for which you think you might qualify. And the next time you see each of the persons to whom you'd turn, say, in an offhand manner, "Should I ever be fired, Joe, you'd be one of the first people I'd turn to for a new job." Keep this list handy at all times. Up-date it every time you meet someone who might help you get a job.

So You'll Have to Relocate

Most engineers and scientists do, sooner or later. For example, one electronics engineer moved five times in six years, owned five different houses in five cities during this hectic period of his career. If you have to relocate, hire a good van service. Don't try to save a few dollars by doing the moving yourself, unless the move is short—say from Los Angeles to Newport Beach, or some similar distance. For anything over 100 miles you need a competent van service. For a helpful booklet on steps to follow in moving your belongings from one place to another, write any large van service.

Your Dream Job

Most engineers and scientists dream of a perfect job. Such a position usually offers more than adequate compensation, liberal fringe benefits like stock options (perhaps even a "piece of the company"), complete intellectual and physical freedom, an unlimited expense account, etc. Since such positions are rare, they're usually called dream jobs. Is the man who dreams of such a job a fool? Let's see.

J. J., a research physicist with a West Coast transistor manufacturer, was thirty-eight years old when he conceived his dream job. After thinking about the job for several freeway trips to and from his transistor job, he sat down and wrote out what this dream job would offer him. Here are his specifications, somewhat condensed to save space: job title: vice president of research; job location: San Francisco, California; minimum annual salary: $25,000; minimum annual profit sharing: 20 per cent of annual salary, or $5,000; other fringe benefits: stock gifts equal in value to one-half the minimum annual profit sharing, hospitalization, and pension to be paid by company; working hours: freedom to come and go at any time. This specification detailed a number of other items related to size of office, type of furnishings and equipment, secretarial help, etc.

J. J. finished his specification, laughed about it a little, and filed it in his desk at home. A few weeks later a friend asked J. J., "Just what would it take for you to switch jobs?"

"I know exactly what I want," J. J. said. "I'll drop my job spec in the mail."

He did. Within a month he had a new job. His new employer met every term of J. J.'s specification. J. J. was just the man the employer needed, and the employer had the dream job J. J. was seeking.

So you see, preparing a dream-job specification can pay off. In fact, preparation of such a spec may show you that your present job *is* your dream job.

REFERENCES

[1] "Engineers deserve more recognition," *Space/Aeronautics*, October, 1961.

[2] I. Hirsch and W. J. Oakes, Jr., "Increasing the Productivity of Scientists," *Harvard Business Review*, March–April, 1958.

[3] "Engineers and Scientists in Industry Feel Their Talents Are Misused," *Business Week*, Oct. 24, 1959.

[4] T. K. Kelly, "Changing Jobs in Today's Turbulent Market," *Chemical Engineering*, July, 1957.

[5] P. J. Brennan, "What It Takes to Get That New Job," *Chemical Engineering*, Feb. 20, 1961.

[6] "Staffing Round-the-world Jobs," *Business Week*, Oct. 7, 1961.

[7] D. A. Moffitt, "Maverick Managers: Individualist Displaces the Organization Man in Many Corporations," *The Wall Street Journal*, Nov. 22, 1961.

HELPFUL READING

Dill, W. R., T. L. Hilton, and W. R. Reitman, *The New Managers: Patterns of Behavior and Development*, Prentice-Hall, Inc., Englewood Cliffs, N.J., 1962.

Eldot, L. D., *Getting and Holding Your Executive Position*, Prentice-Hall, Inc., Englewood Cliffs, N.J., 1960.

Forrester, Gertrude, *Occupational Literature*, The H. W. Wilson Company, New York, 1954.

Glanvelle, J. L., *Modern Vocational Trends Handbook*, World Trade Academy Press, New York, 1957.

Haldane, Bernard, *How to Make a Habit of Success*, Prentice-Hall, Inc., Englewood Cliffs, N.J., 1960.

Pitt, G. A., *The Twenty-Minute Lifetime: A Guide to Career Planning*, Prentice-Hall, Inc., Englewood Cliffs, N.J., 1959.

Rood, Allan, *Job Strategy: Preparing for Effective Placement in Business and Industry*, McGraw-Hill Book Company, Inc., New York, 1961.

PART 2

Learn the Art of Management

13

Get Along with Your Boss

> Know your boss; he's the second most important person in your life. And if you wish to succeed, let him know you by your actions.
>
> ANONYMOUS

No matter how much you achieve in engineering or science, chances are good that you'll always have a boss. For everyone, from the chairman of the board, president, vice president, and on down, has someone to answer to. Recognize this fact as early in your career as possible, and you'll achieve more. Why? Because your output is greater when you work *with* your boss instead of *against* him. Let's see how you can work more effectively with any boss.

A Positive Attitude Wins

Engineers and scientists, trained in the scientific method, sometimes adopt a "show me" attitude in all their dealings with the boss. As a result the boss finds these men obstinate and unproductive. He dreads giving an order because each directive leads to an argument. The boss hasn't the time, energy, or desire to argue over every order he gives. Instead, he wants action—and early completion of the task. When you give him action and results he is grateful. In general, a boss likes a willing, striving subordinate who regularly gives average results. He dislikes the argumentative, obstinate worker, even though this man may have an occasional flash of ingenuity. The boss can depend on the willing man—he never knows when the argumentative man will perform the assigned task.

Look at your boss—or any boss—today. The chances are excellent that he is (*a*) somewhat harried, (*b*) overloaded with work, (*c*) under pressure from three or more different sources, and (*d*) constantly trying to do a good job. To get along well with him you must recognize these conditions and try to help him in every way possible. You must be positive in your outlook, seeking to assist, instead of thwarting. Here are 19 hints showing how a positive attitude in your dealings with your boss will help you achieve more.

19 Hints for Getting Along with the Boss

1. *Don't run to him with every little crisis.* Make some decisions yourself. *Your* boss won't let you make decisions? That's nonsense. Review your areas of responsibility. Make decisions within these areas. If you haven't the right to make decisions in these areas, see the boss. Tell him you want to take a load off his shoulders by functioning as his helper—not his errand boy. Show him how much you can help him, and he'll be delighted to assign you the responsibilities you request.

2. *Respect his time.* Don't dally in his office—say what you must and leave. Limiting your conversation will keep the talk focused on essentials. This way there's less chance of saying the wrong thing or of wasting time on trivia. Respect your superior's time, and he'll respect yours. You'll both accomplish more in greater harmony.

3. *Stay away from your boss when he's busy or has something important on his mind.* Study your man; get to know when he's preoccupied. You can tell from the expression on his face, the tempo of his walk, or the sound of his voice. Interrupt at these times and you may receive a fast, unpleasant brush-off. So wait, if at all possible, until his mind is clear, and he can give you full attention. Where you have no alternative but to interrupt, be certain that you are as concise as possible and leave his office as soon as you are finished.

4. *Use memos and notes to save his time.* The door is always open to a memo or note. Written communications save time, serve as

records, and get results. Some executives prefer written documents because they can be read anywhere—on a train, plane, at home, or in the office. By judicious use of memos and notes you can often penetrate the thickest mahogany barrier between yourself and your boss.

5. *Act more—talk less.* You're paid to do, achieve, deliver. Take immediate steps to show results. Limit your conversation with your boss to essentials—unless he asks for more information. Keeping your talk under control will give you more time to complete your tasks. Since this is what you're paid for, your worth will rise in the boss's eyes.

6. *Help—don't hinder.* Every intelligent boss will respect and admire your opinions. But when he gives an order, he expects you to carry it out—regardless of your personal opinions. So seek ways to help him. One of the most important ways of helping is that of doing what you're told to do promptly and efficiently. Never thwart your boss—help him. This is exactly what you expect and want from the people you supervise.

7. *Develop alternate solutions to problems.* Never dash, scatterbrained, into the boss' office, expecting him to have all the answers. Study the problem, situation, or facts in advance. Develop alternate solutions, methods, procedures. Make notes about these so you won't forget should you become flustered during your talk with the boss. List the advantages and disadvantages of each alternate and be ready to cite these when necessary. Every alternative you suggest may be turned down, but the help and stimulation you provide will be welcomed by the boss. Even though he turns down every alternative, the solution he suggests will probably be a combination of several of your ideas.

8. *Steer clear of your personal life, unless asked.* The boss is probably interested in you, your family, your hobbies. And he wants to hear about them—but only when he asks. So don't volunteer to tell him about your handsome son or talented daughter until after the boss mentions his children. Be careful not to engage in one-upmanship over children, golf scores, club memberships, etc. *No boss—regardless of how generous or open-minded he may be—wants to be topped by his employees.* This is a human desire. As an engineer or scientist, keep that fact in mind at all times.

9. *Catch him at odd moments.* There are many places to talk to

him, other than his office. Catch him in the hall, elevator—even the company cafeteria. His mind is more likely to be free at such times and he can give a quick answer to a short question. Confine such questions to routine matters. Your boss will resent having confidential matters discussed in public. Be sure he's not talking to someone else when you ask him a question or tell him of a new development. He might resent your intrusion.

10. *Get to know your boss.* Your supervisor can be, next to your wife, the most important person in your life. So learn what he likes in the way of reports, data, facts, etc. Presenting information in terms he favors can build a valuable rapport. He'll be able to analyze your data faster. And he'll think more highly of you. You'll both accomplish more in less time.

11. *Never complain to your boss.* He gets enough complaints from other sources—his boss, his secretary, foremen, supervisors, etc. Translate your complaints into suggestions. Any suggestion is much easier to listen to than a complaint. Besides, a complaint is negative while a suggestion is positive. No boss likes a complainer, but almost every boss welcomes helpful suggestions. Smile and suggest instead of complaining and crying.

12. *Avoid bullying.* Some engineers and scientists take devilish glee in reminding the boss of an imminent deadline, forgotten report, or other unfinished task. They like to see the boss in trouble. You never gain by embarrassing anyone. Bullying only builds resentment. So when you feel a strain of perverseness coming on, remember hint (6) *Help—don't hinder.* A helpful attitude can build cooperation—bullying can lose you your job.

13. *Don't look to your boss for pity.* This is one emotion every experienced supervisor avoids. Pity doesn't help you or the boss. So instead of seeking pity, search for ways of solving your problem. Pity holds you back, delays decisions, wastes time. Substitute positive action for pity and watch your problems disappear.

14. *Search for the best communication channels.* Keep trying until you learn how to get through most effectively to your boss. Perhaps he's the memo type—a man who prefers having every important item in writing. Or he may be the meeting type—a man who solves his problems by group effort. Once you've found his preferred methods, use them whenever possible. Doing so will ensure more rapid and effective communication.

15. *Help keep your boss informed on what's new.* Tell him about new developments, trends, procedures. Express your news in short, terse form. You'll find that a competent supervisor will be aware of some of the items you tell him about. So don't be windy. Instead, adopt a friendly, helpful tone that gets information across as quickly as possible. And never, never, allow a superior tone to creep into your voice. You'll only make the boss feel backward. This may bias him against you for a long time. Engineers and scientists, constantly in search of new knowledge about their field, sometimes become cocky and overbearing when they discover important facts. Avoid this feeling—it will never help to keep you on friendly terms with the boss.

16. *Be profit- and time-conscious.* Almost every supervisor is concerned with making a profit for his organization—either directly or indirectly. So gear your thinking to profits. Just because you're an engineer or scientist doesn't mean that you can sneer at profits. For profit, in industry, is what provides the capital that enables a firm to hire you. Without profit your job wouldn't exist for long. Get over the widely held concept of many engineers and scientists that profits are "dirty" because they concern money. This is a juvenile attitude that holds back capable engineers and scientists who might otherwise move ahead and achieve more. Profits are here to stay, and the sooner you recognize this basic fact of economics, the quicker you will achieve greater rapport with your boss.

Time is an integral part of every profit-producing venture. Remember this on all occasions. Save a few hours here and a few hours there and profits will rise. So seek ways to save time for your boss and everyone in his area of responsibility. Your reputation will soar along with your achievements. Remember—your boss *has* to be profit- and time-conscious. So must you, if you ever expect to take over his job.

17. *Show enthusiasm for your job.* Your boss will excuse many more mistakes, if you have a basic enthusiasm for your job. Lack of enthusiasm will make you a bore to have around, will give a drab tone to your work. Take joy in what you do and others associated with you will improve the quantity and quality of their work. Ask any supervisor whether he likes an enthusiastic employee and the answer will invariably be yes. So drop the pose of bored sophistication which so many engineers and scientists adopt. Let your interest

and pleasure in your work show through. You'll benefit more than you ever thought possible.

18. *Be quick, decisive.* Learn to solve problems more rapidly. Present your solution as soon as possible. No boss likes a hesitant, fumbling employee who is indecisive and unsure of himself. Some engineers and scientists hide behind a facade of deep knowledge and complex technical jargon. This, they hope, will save them from making a decision. Avoid such an attitude—it defeats whatever good you accomplish and is one of the reasons why managers sometimes call engineers "a strange breed." Sure, some of your problems are difficult from a technical standpoint. But this is not enough reason for throwing up a protective smoke screen. So instead of ducking behind a mishmash of technical mumblings, try to give an approximate answer, a guess, a trend. Your boss will realize that you're "taking a flyer" and won't hold you to exact results. But he will appreciate your effort. What's more, he'll look on you as a quick, decisive, and dependable engineer or scientist who rates a promotion.

19. *Don't hide your abilities from your boss.* Show him, by positive action, what you can accomplish. Don't wait to be told to do routine jobs—do them when you see them waiting. Independent action, well-planned and executed, will be noticed by even the most unobserving boss. He'll remember your initiative when he has responsible tasks to assign. You'll achieve more and get more notice when you use positive results to show your abilities. Remember: One job well-done is worth hours of conferences in which you tell the boss how good you are.

Using the 19 Hints

May any of these 19 hints be ignored? Certainly, if conditions make it necessary for you to do so. These hints are offered as general guides. You may find that one or more of them is not applicable to your particular situation.

Some engineers and scientists are more human relations conscious than others. These men have an innate ability to get along with their boss. Other less fortunate technical and scientific personnel are awkward in their dealings with their supervisors. If you're one of these people, study the 19 hints carefully, and apply those you believe will help you. Carefully observe the results. You will probably find that you and your boss accomplish more with less friction.

All of us, regardless of our positions in the engineering and scientific fields, will always have a supervisor. Since poor relations with the boss hinder ourselves, the organization, and the boss, everyone gains when we take steps to improve the situation. So resolve today to be a dependable, helpful, and self-starting employee. Learning to manage yourself will put you in line for promotion to a spot where you manage others. You are certain to achieve more in your chosen profession when you learn how to get along with your boss.

HELPFUL READING

Davis, Keith, *Human Relations at Work,* 2d ed., McGraw-Hill Book Company, Inc., New York, 1962.

Dyer, Frederick, Ross Evans, and Dale Lovell, *Putting Yourself Over in Business,* Prentice-Hall, Inc., Englewood Cliffs, N.J., 1957.

Shidle, N. G., *Instincts in Action,* Publishers Printing Company, New York, 1962.

Uris, Auren, *The Efficient Executive,* McGraw-Hill Book Company, Inc., New York, 1957.

——— "Yes, You *Can* Get the Boss's Ear," *Factory Management and Maintenance,* April, 1959.

14

Think and Act Like an Executive

> Great thoughts reduced to practice become great acts.
>
> WILLIAM HAZLITT

Let's face it. A top executive in any branch of engineering or science is a different breed of man.[1] Compared with men in the lower echelons of management, he's smarter than most. He works harder but knows how to relax better. He has much more innate understanding of a variety of problems and effective techniques for solving them. Today's typical successful engineering or scientific executive has an irresistible urge to be at the top, to earn more money, to make himself heard, to win acceptance for his ideas—to contribute more to his profession. He wants, if at all possible, to dominate his specialty.

Above all he knows how to get other people to implement his ideas and ambitions. He has, in short, that everlastingly elusive something called leadership. As John Erskine said, "In the simplest terms, a leader is one who knows where he wants to go, and gets up, and goes."

But all the leadership, ambition, and talent go for naught unless the technically trained executive is going in the right direction—or knows which direction to choose. Many a jet-propelled wonder boy has come over the horizon only to disappear in a blast of smoke as he went off target.

The Successful Executive

Regardless of his background the successful executive makes a career of studying each job he holds to determine what its objectives are. Oliver Cromwell said, "No one rises so high as he who knows where he is going." This applies to every executive today. He achieves continuing success by developing himself to meet the job requirements. Later chapters in this book show you how you can develop your particular talents to meet specific job needs. The present chapter gives you many useful hints that will serve you in any executive position you may hold now or in the future.

Four Common Denominators. The common denominators in an executive's job determine the kind of man who can handle it. Four important characteristics of such a man are:

1. *He knows his capabilities—and limitations.* He's a harsh judge of his own abilities. Only when he sees himself objectively can he reinforce his personality where it is weak. He rounds out his personal inventory of skills by surrounding himself with competent specialists.

2. *He regards himself as a professional administrator.* He doesn't worry about a minor lack of knowledge of the technical aspects of his job. He avoids the pressure of trying to know everything about everything. He understands the principles of delegation and uses them to make his job easier.

3. *He's a student of human nature.* This is not because he loves everybody. Probably he doesn't. But because he knows that shrewd judgment in estimating peoples' loyalty, integrity, and knowledge will enable him to get more out of them. Even the toughest, roughest industrialists are superior in their ability to inspire cooperation from their associates. Perhaps they're successful because they don't deal in any flimflam about human relations for the good of the soul. They have a feeling that they know what's good for people.

4. *He knows how to think straight.* People tend to think of the successful executive as the classic movie or TV type of administrator —as a fellow who makes every decision easily and without hesita-

tion. The real executive, regardless of his field, seldom fits this pattern. He isn't worried if subordinates look critically at him when he refuses to make a decision until a situation absolutely demands it.

"Positive" procrastination *can* be a method of delegation. It forces others in your organization to commit themselves to stick their necks out—in short, to develop. Taken out of context, this may sound like heresy. But it's an example of an executive's ability to find out what a problem is all about and be clear-headed in the action he chooses. This is one way of saying he uses a normal thought process.

The Normal Thought Process

Centuries ago the Greeks determined what it was that made some people smarter than others. They worked out a formula but somehow it was forgotten until John Dewey rediscovered it about fifty years ago. You probably recognize its five steps:

1. ***Gather all the facts.*** The biggest mistake is to make a decision without having enough information. This is a big temptation when you're trying to act in a hurry.
2. ***Determine the real problem.*** Smart executives concentrate their effort here, using specialists to work out the answers.
3. ***Analyze the problem.*** The easiest approach is to compare the benefits you'll get from solving the problem with the obstacles that stand in the way of a solution. This technique will often reduce the problem to smaller ones that are easier to solve.
4. ***List possible solutions.*** For each obstacle there *has* to be at least one possible solution. Now you must decide how to overcome the individual obstacles.
5. ***Develop a plan of action.*** The most obvious move, but often overlooked or evaded, is deciding *what* to do. You can't stop there, though. You must also determine *who* is to do it, *how, when,* and *where.*

Thinking and Acting

You cannot separate thinking like an executive from acting like an executive. One is father to the other. Straight thinking starts as an inactive mental process, ends as an action-taking decision. As an

What's your average? Check these symptoms on the scoreboard

BETTER CHECK NOW

	ALWAYS?	MOST OF THE TIME?	SOME TIME ?	NOT AT ALL?
1. You don't feel on top of your job. You fall a little further behind each day.				
2. Major decisions are almost impossible to make. Minor ones nibble away at your confidence.				
3. You daydream, especially in time of crisis. Or you get to thinking, "If only I had known that."				
4. When things go wrong, the first thing you do is look for a "fall guy."				
5. You have a feeling your boss "sees through you." You're always covering up.				
6. You're sure the only way to handle strong subordinates is to get rid of them.				
7. You can't seem to get along with your associates. Or you're sure they're plotting against you.				
8. It has been a long time since you've admitted a failure as your own responsibility. You shift emphasis elsewhere.				
9. You just can't keep your temper. You're alternately blowing your top and trying to make amends.				
10. You respond to the urge to get away from the job when pressures are high. Or, for no known reason you get sick at crisis time. Or you can't get through the day or night without a drink---or several drinks.				

FIG. 14.1. Check your thinking habits and score yourself. Take action to improve, if your score is low. (*Lester Bittel and Factory Management and Maintenance*)

executive you begin by using the normal thought process we discussed above. To this you add your educated guesses and your good hunches. These three combined equal executive performance. That's what you're judged on.

Take a moment now to check your present thinking habits, using the checklist in Figure 14.1. Rate yourself for each question, using the space provided. Don't let a poor score get you down. Remember, no one is perfect. Even the best executives err now and then. Still, if most of your checks fall in the "always" or "most of the time" columns, you have a real job ahead of you to improve your thinking. But take courage, help is near at hand to assist you in supplementing the normal thought process.

To help you supplement the normal thought process, here are 14 ideas for thought and action that smart executives use. You can use them, too, to increase your personal effectiveness on the job.

The 14 Ideas

Study these ideas *now*. Start using them *today*. For, as the Chinese sage said, "Even a journey of 1,000 miles starts with a single step."

1. ***Avoid crisis decisions.*** Don't try to dash about, "putting out fires." This is the mark of an amateur. This *modus operandi* may be all right for the supercharged person with an overactive inferiority complex. But for most executives in engineering and science it's poison.

Many a top flight man has come into a crisis-ridden department as a troubleshooter. By careful planning he gets things running so smoothly that soon there are hardly any decisions to make. Then he pulls out—to a better job somewhere else in the organization—and a lesser man can take over. If you develop the habit of looking ahead and making decisions at a comfortable pace, you can become an accomplished executive.

2. ***Distinguish big from little problems.*** Some people never know when opportunity knocks—or when a critical decision is at hand. Weigh the benefits you'll get, or the headaches you'll avoid, by taking corrective action. This will show you when an all-out effort is called for. To do this you must develop poise and flexibility and become a man who can sense a critical situation and rise to it. Also, the good executive is the kind of man who can stay in action all day every day.

3. ***Organize for delegation.*** Don't overlook the importance of good organization for delegating routine tasks. In its simplest form this can mean looking over the job responsibilities in your group to be sure someone is regularly assigned to every one of them and to make sure that each person has authority to carry out his own responsibilities. Perhaps you wear six hats and your assistant five. You needn't sign a contract for each hat, but it certainly helps to chart out what you do and what you delegate. And remember: Delegation can be in little things—such as allowing your assistant to decide how his new office should be arranged, what kind of desk he'll have, and how he'll handle the details of his job.

4. ***Rely on policy.*** It will settle many routine decisions for you. Remember—it's a waste of time and effort to make the same decisions more than once. Reduce repetitive decisions to policy. It may require much thought and effort to develop policies, but they save you from troubleshooting later on. Be sure to inform your subordinates of all policy decisions—and what you expect from each person under your jurisdiction.

5. ***Rely on others for counsel and advice.*** Today it's a sign of strength rather than weakness not to know all the answers. It's

smart to solicit guidance from your associates and superiors. "Experience," Carlyle said, "is the best of schoolmasters; only the school-fees are heavy." Seeking advice from others is good human relations, too. It allows for participation. It's flattering to those whose opinions are asked.

6. *Use good communications to anticipate trouble.* You don't want a flock of gossips and talebearers working for you. But you do want to make a regular practice of talking with your staff, formally and informally. If they know what might give you trouble, they can report danger signals to you. It's valuable to have a man who works for you say: "Stop me if you've heard this. Maybe you aren't interested. But I was in the purchasing department today and heard they're changing from our regular supplier to an outfit that has welched on delivery dates with us before."

7. *Pace your thinking efforts.* Don't face each problem as if it were a major crisis. Rise to the tough ones, but learn to regard most decisions as routine. Make your decisions calmly and confidently. "Think calmly and you'll be surprised by the clarity of your thoughts and ideas," said Howard Chase.

8. *Establish a control system.* The key to decisive action for many successful executives is a control system that alerts them when critical factors are failing to meet expectations. You may say, "What! more paper work?" Or, "I can keep it in my head." Controls needn't be complicated. Make it easy for yourself. Take a look at your job. List its five or ten most critical items—like costs, meeting schedule dates, turnover, indirect labor costs, scrap totals, etc. Circle your calendar to remind you to check them once a week, or whenever your experience shows they tend to change.

9. *Set up guides for handling your human relations problems.* Working too strenuously at human relations takes much time and energy. Try to develop the habit of dealing with people in a way that is psychologically sound. (See Chapter 16 for a variety of useful hints.) Avoid being a "parlor psychiatrist." Don't try to outsmart and outguess people—this only complicates a situation. If your contact with people is direct, sincere, and intelligent, you'll keep your human relations reasonably trouble-free.

10. *Develop the habit of concentration.* Devote full attention to the problem at hand. Don't allow your mind to shift back and forth from one problem to another. When this happens, neither problem receives the attention it should. Rid yourself of routine details by assigning these tasks to others. Use reminder notes and your call-up file to help your memory. Once you take these steps, you will be better able to concentrate on the problem facing you. With improved concentration comes greater facility to solve all problems arising in your work.

11. *Take your time.* Some tasks require immediate action. Others can wait a day or two, or longer. Never put off a decision because you feel lazy. Ben Franklin had the right idea when he advised, "Never leave that till tomorrow which you can do today." Putting off decisions makes them pile up. However, don't rush a decision—take time to get more information, to reflect, and to check yourself. Then your percentage of correct decisions will be higher.

12. *Don't expect perfect results.* Be smart—know what kind of results to expect from the decisions you make. Study the situation so you know your chances of being right—or wrong. In making any decision it's difficult to be completely right—to satisfy every person or condition involved. Understand the odds and govern your actions accordingly

13. *Draw a picture of your problem.* Nothing helps you visualize a problem so much as a diagram. Just as visual aids help others to see an idea clearly, so your own sketches help you see your problem and work out solutions. Use the sketch to help keep your mind from being cluttered with related but nonessential ideas. On paper you can see your main ideas more clearly. And less significant ideas can be shown in their proper perspective and relation to the main idea. Try Ben Franklin's method, which he described in a letter to Joseph Priestley, the British chemist who made so many important contributions to the study of gases. When Lord Shelburne offered Priestley a position as his librarian and literary companion, Priestly sought Franklin's advice. Franklin wrote:

> *London, September 19, 1772*
>
> Dear Sir:
>
> In the affair of so much importance to you wherein you ask my advice, I cannot, for want of sufficient premises, advise you *what* to determine, but if you please I will tell you *how*. When these difficult cases occur, they are difficult chiefly because while we have them under consideration, all the reasons *pro* and *con* are not present to the mind at the same time; but sometimes one set present them-

> selves, and at other times another, the first being out of sight. Hence the various purposes or inclinations that alternately prevail, and the uncertainty that perplexes us.
>
> To get over this, my way is to divide half a sheet of paper by a line into two columns; writing over the one *Pro,* and over the other *Con.* Then during three or fours days' consideration I put down under the different heads short hints of the different motives that at different times occur to me, *for* or *against* the measure. When I have thus got them all together in one view, I endeavour to estimate their respective weights; and where I find two (one on each side) that seem equal, I strike them both out. If I find a reason *pro* equal to some two reasons *con,* I strike out the three. If I judge some two reasons *con* equal to some three reasons *pro,* I strike out the five; and thus proceeding I find at length where the balance lies and if after a day or two of further consideration, nothing new that is of importance occurs on either side I come to a determination accordingly. And though the weight of reasons cannot be taken with the precision of algebraic quantities, yet when each is thus considered separately and comparatively, and the whole lies before me, I think I can judge better, and am less likely to make a rash step; and in fact I have found great advantage from this kind of equation in what may be called *moral* or *prudential algebra*. . . .

Priestley accepted the position in 1772. In the same year he was chosen a Foreign Associate of the French Academy of Sciences. Whether Priestley used Franklin's "way" we do not know. But this method is as useful today as it was 200 years ago. Certainly if it was good enough for Priestley, who associated with Boulton, Watt, Lavoisier, and Erasmus Darwin (Charles's grandfather and a scientist in his own right), it is worthy of our consideration today. You may not accept every step in this method, but it can give you many ideas for developing your own decision-making aids.

14. *Have alternate solutions available.* Situations, facts, and conditions keep changing. So you're wise to have at least two answers for each problem. Keep your second solution in reserve to use if conditions change—or if you were wrong in the first place.

You Can Think Better

Some engineers and scientists strongly resist any change in their thinking habits. They prefer to stay within the rigid confines of their specialty. While this is satisfactory for a man who wants to specialize, it won't do for those seeking broader achievements. If you want to develop a broader career, begin today to think and act like an executive. For no matter how unimportant your duties may be, a more creative approach will build your skills. Then when your chance comes, you will be ready for it. For "to improve the golden moment of opportunity, and catch the good that is within our reach, is the great art of life," is as true today as when Samuel Johnson first said it over 200 years ago.

REFERENCES

[1] Lester R. Bittel, "How to Think- and Act-Like an Executive," *Factory Management and Maintenance,* December, 1954.

HELPFUL READING

George, C. S., *Management in Industry,* Prentice-Hall, Inc., Englewood Cliffs, N.J., 1959.

Hepner, H. W., *Perceptive Management and Supervision: Insights for Working with People,* Prentice-Hall, Inc., Englewood Cliffs, N.J., 1961.

Herron, L. W., *Executive Action Simulation,* Prentice-Hall, Inc., Englewood Cliffs, N.J., 1960.

Mahoney, T. A., *Building the Executive Team: A Guide to Management Development,* Prentice-Hall, Inc., Englewood Cliffs, N.J., 1961.

Planty, E. G., and J. T. Freeston, *Developing Management Ability,* The Ronald Press Company, New York, 1954.

Uris, Auren, *The Efficient Executive,* McGraw-Hill Book Company, Inc., New York, 1957.

[illegible] While this [illegible] for [illegible] who want to [illegible] for those [illegible] [illegible] the [illegible] of [illegible] [illegible] how important your [illegible] [illegible] approach will build your skill. Then when [illegible] you will be ready for it. [illegible] opportunities and catch the [illegible] of life" is as true today as when Samuel Johnson [illegible] said it 200 years ago.

REFERENCES

[illegible]

HELPFUL READING

[illegible] Prentice-Hall, Inc., Englewood Cliffs, N.J., [illegible]

[illegible] Prentice-Hall, Inc., Englewood Cliffs, N.J., [illegible]

[illegible] Prentice-Hall, Inc., Englewood Cliffs, N.J.

[illegible] Company, New York, [illegible]

[illegible] McGraw-Hill Book Company, Inc., New York, [illegible]

15

Judge People Better

> A man must have his faults.
>
> PETRONIUS

> We are always sizing up others, so let us learn how to do a better job of it.
>
> D. A. LAIRD & E. C. LAIRD

To contribute more to your field you must, in general, work closely with many people. As your responsibilities grow, so too will your need for accurately judging or sizing up people. The ability to correctly judge people enables you to choose able assistants, weed out incompetents, assign capable men to specific tasks, and to build an efficient, skillful team in your department or division. Knowing how to judge people can mean the difference between an outstanding career and a mediocre one.

Liking People Builds Judging Skills

One of the biggest problems some engineers and scientists face in judging people is their lack of interest in anyone outside their own field. And even if an engineer is interested in people outside his own field, he may persist in judging these people by weak, ineffectual standards—like color of hair, school attended, club memberships, etc.

To effectively judge people you must develop an interest in and liking for most people you meet. For if you do not, there is little chance of acquiring an understanding of why people act the way they do. If you do not understand people, then your chances of motivating them are slim. And while you may achieve limited goals without the help of people, your chances for outstanding achievement are limited unless you can judge most people accurately most of the time. For it is the help, knowledge, and skill of other people that will enable you to increase the importance of your achievements in your chosen field.

Rid Yourself of Prejudices

Almost everyone has prejudices of some kind. Certain prejudices are harmless—others can crimp your career more than you ever thought possible. For as Hazlitt said, "Prejudice is the child of ignorance." Prejudice can make you turn a capable man away or lead you to choose an incompetent for an important task.

How can you discover your prejudices? [1] It isn't easy—most people, when asked what their prejudices are, will say, "None!" Yet few of us are free of prejudice. Only when we dig deeply into our beliefs do we realize that we are mildly or seriously prejudiced about certain things, people, beliefs, etc. Stop for a moment and ask yourself what you find most irritating in the people you meet. Does your list include:

Slovenly clothes	Ivy-league cut clothing
Loud talk	Unpolished shoes
New York Yankee fans	Southern accent
Bearded chin	Business school graduates
Dark complexion	Medical doctors
Horn-rimmed glasses	Slouching posture

If your list includes any of these or similar items, you are somewhat prejudiced because none of these, of themselves, should be sufficient to turn you against a man. For after all, every man is entitled to his opinions, beliefs, enthusiasms, and minor follies. When you want a job done, it is the big aspects of a man's character that count. So learn to tolerate minor faults that crop up now and then.

How can you rid yourself of prejudices? There are two direct steps you can take: (1) Get closer to and more friendly with as many people as possible; and (2) spend as much time as you can with people. Let's see how and why each step will help you say goodbye to useless prejudices.

1. Pick irritating people as your new friends and associates. By doing so you may find that the bold, brash individual you originally disliked is actually an extremely competent and level-headed engineer or scientist. His boldness and brashness may actually be assets to your organization or to the advancement of his personal goals. Your prejudice against boldness and brashness will disappear as you become more friendly with such a man and see what he achieves in his career.

By associating with as many people as possible you will gradually lose many of your unreasonable and unreasoning prejudices. Thus, bearded scientists will no longer provoke you to unkind thoughts; you'll tolerate bow-tied engineers; you may not root for the same team as New York Yankee fans, but you'll pass up condemning a man for this leaning. Experience with "people in the round," which many engineers and scientists avoid, will teach you that most of your prejudices are misleading. You'll also learn that you can't "type" people. For the moment you type a man as an intellectual, you'll find that he has many nonintellectual interests and abilities—for instance boxing or wrestling.

2. Spend as much time as you can with people. Don't hide from them—you gain nothing when you avoid people. Being with people much of the day gives you a chance to observe them under a variety of circumstances. You'll learn about their personal problems, hopes, beliefs, desires. You will slowly see that basically most men are honest and trustworthy. Their little differences and foibles will become interesting instead of irritating. Where a year ago you disliked all red-headed engineers, you'll now find two of your best friends have this color hair—but you never notice it.

How can you spend more time with people? Go to lunch with them. Become active in your professional society (Chapter 26). Use odd moments during business trips (Chapter 10). Work with volunteer groups in your organization for the benefit of a good cause like the Red Cross, etc. Join a team in your organization—bowling, softball, golf. Go where people are, and you will be certain to spend more time with them.

Try as much as possible to see other people as they are. Do not shun a man because he fails to measure up to certain fixed ideas—*prejudices*—you have. Instead, seek out the person you find unattractive. Spending time with him will often reveal positive characteristics you never thought he had. Once this happens, your prejudices will begin to grow weaker.

You'll never rid yourself of every prejudice. Human nature is such that almost everyone holds a few prejudices. But as long as you keep trying to unburden yourself of unreasonable prejudices, you will grow in maturity and capability. For as Lowell said, "The foolish and the dead alone never change their opinion."

Once you've reached the state where you have very few prejudices, recognize those remaining. Then, when you're judging a person, take these prejudices into account. If you dislike bearded scientists and would hire a new man, except for this one drawback, forget his beard and hire him. You will probably find that he's the best addition you've made to your staff in many years.

Eight Guides for Judging People

Here are eight guides you can use when sizing up a man. While these guides aren't infallible, you'll find they are correct most of the time with most of the people you meet. Begin using them today and see how your ability to judge people grows.

1. *Allow him to talk.* Don't fill the time you spend with a man with your talk. Get him talking about himself and his problems. For when he talks freely, he reveals far more about himself than when he answers a series of questions you put to him. Use wise interview techniques (Chapter 21) and you'll learn more in less time.

2. *Listen attentively.* Don't allow your mind to wander. *Hear* what he says (Chapter 20). Note how he talks—does he whine, boast, threaten, or mumble? Each habit may be a key to his main personality traits. A whiner is likely to be an inefficient time waster who upsets his associates. Braggarts seldom produce what they claim they can or will. The bully threatens but doesn't often contribute much. And the man who mumbles may be too introverted or bashful to complete much more than routine tasks. So listen while he talks—you'll benefit more than you ever thought possible.

3. *Observe him carefully while he talks.* Start watching as soon as you see him. Observe his walk—is he quick and confident or slow and doubtful? Begin your detailed observation with his hair and note everything about him down to his shoes. Is his hair neatly trimmed and combed? What general impression do his face, eyes, and mouth convey? Does he look you in the eye or does he avoid a direct glance? Are his teeth cared for? Is his voice calm, well-controlled? And what about his English? Does he speak grammatically? Are clichés a frequent part of his speech?

What about his clothing? Is it appropriate for the meeting? Is his suit loud, poorly fitted, or unpressed? Are his shoes polished?

Watch his hands while he talks. Does he emphasize his remarks with strong gestures? Are his nails clean and trimmed? Does he twiddle his fingers, wring his hands, or put one hand over his mouth while he talks?

Why should you observe these items about a man? Steinmetz was a hunchback who smoked big cigars; Einstein never won a prize for being the best-dressed scientist. True—but men of their caliber are rare. For the average engineer or scientist the items listed above can help you gain a better concept of his true character. Thus, the sloppy man—unpressed suit, dirty fingernails, untrimmed hair—may have undesirable thinking habits. If you're considering him for a task in which he'll have to meet the public, he may make a poor impression on prospective customers.

Never reject a man solely on the basis of appearance. Judging a man accurately requires that you weigh what you see, hear, and interpret. Unpolished shoes may be an oversight. Uncared-for teeth may, however, indicate a man does not budget his expenditures to include his most important possession—his body.

4. *Seek the big accomplishments in his career.* Learn how he has proven himself. Does he have a long, steady, satisfactory employment record with one organization? Or does he hop from job to job for little reason? Is he active in his professional society? Or does he let the other guy do it all? What about friends? Does he have many? Does he have any hobbies? Has he ever published any professional articles, papers, or books?

Every man worth anything to his profession will have accomplishments in areas that interest him. He will be proud of these accomplishments and, in most instances, will be willing to talk about them.

Certainly the man who has established a steady employment record in the time available to him, who has contributed to his professional society, who has many friends and a hobby or two is a likely candidate for any task you have in mind. For a man like this has a focus in his life. He is in that group about which Virgil said, "They are able because they think they are able."

In general, the man with a job-hopping employment record, lack of interest in his professional society, and no hobby, lacks overall direction in his life. While he may accomplish miracles for you, it's more likely that he'll turn in a mediocre performance. Ask yourself this question: Who's more likely to do a good job, the man who spends his spare time in professional society work or the man who puts most of his spare time into supporting the owner of the local pub?

5. *Be alert to topics he avoids.* We all avoid that which is painful. If a man constantly finds faults in other people, organizations, equipment—if he avoids the good that is in everything—beware. He'll be finding fault in you and your organization soon. Unfortunately, there is a tendency in some engineers and scientists to ignore the positive and seek the negative. Such an attitude almost always leads to a fall-off in morale and efficiency. So be on guard for the man who avoids the good and seeks the bad.

Watch for the man who avoids a direct answer to your question. If he has nothing to hide, he'll answer you. Avoiding the answer shows that he will not be forthright with you. The secretive or sneaky man seldom makes a good associate.

6. *Let a man show you his personality.* Don't build it for him with your guesswork. Listen and observe—most people will tell and show you enough to permit you to form a personality picture of them. Don't be led astray by superficial details—a bald head, curling mustache, southern drawl. Search out his real personality from what you hear and see. Remember what Chesterfield said, "You must look into people as well as at them."

Probably the greatest errors you will make in judging people will be snap decisions as to a man's ability, reliability, and future potential. So avoid the quick judgment of a man and give him time to show you his true worth. Steer clear of typing a personality, i.e., concluding that all bifocal wearers are intellectual or that a New England accent denotes a Harvard graduate. Typing a personality can lead you to misjudge people more often than you judge them correctly.

7. *Keep initial judgments tentative.* Don't form final conclusions until after you've seen a person in action awhile. Tentative judgments allow you to alter your opinion of the person as you see him perform. Once you've acquired enough knowledge of him, you can rate his personality, skills, and potential. Your final judgment is likely to be far more accurate because you've taken time to see the person under a variety of circumstances.

8. *Work at developing your skills.* Learning to judge people is a life-long task. Work at it regularly, and you'll find your efforts interesting and rewarding. You'll also learn that sizing up people by guess isn't half as efficient as the use of intelligence and thought. Try it today and see. Keep in mind at all times La Rochefoucauld's sage observation: "The true way to be deceived is to think oneself more clever than others."

REFERENCES

[1] Donald A. Laird, and Eleanor C. Laird, *Sizing Up People,* McGraw-Hill Book Company, Inc., New York, 1951.

HELPFUL READING

Argyris, Chris, *Personality and Organization: The Conflict Between System and the Individual,* Harper & Brothers, New York, 1957.

Davis, Keith, *Human Relations at Work,* 2d ed., McGraw-Hill Book Company, Inc., New York, 1962.

Fleishman, E. A., *Studies in Personnel and Industrial Psychology,* The Dorsey Press, Inc., Homewood, Ill., 1961.

Gilmer, B., *Industrial Psychology,* McGraw-Hill Book Company, Inc., New York, 1961.

Guilford, J. P., *Personality,* McGraw-Hill Book Company, Inc., New York, 1959.

Laird, Donald A., and Eleanor C. Laird, *Sizing Up People,* McGraw-Hill Book Company, Inc., New York, 1951.

Leavitt, H. J., *Managerial Psychology: An Introduction to Individuals, Pairs, and Groups in Organization,* The University of Chicago Press, Chicago, 1958.

Marrow, A. J., *Making Management Human,* McGraw-Hill Book Company, Inc., New York, 1957.

Palmer, Stuart, *Understanding Other People,* Fawcett Publications, Inc., Greenwich, Conn., 1956.

Spates, T. G., *Human Values Where People Work,* Harper & Brothers, New York, 1960.

16

Human Relations Always Counts

> Success in dealing with other people depends on a sympathetic grasp of the other fellow's point of view.
>
> HENRY FORD

A skill some engineers and scientists deride more than any other is the one popularly called *getting along with people.* Yet this is probably the one skill that is most lacking among today's engineers and scientists. Why is this skill lacking? Possibly because during their college years most engineers and scientists concentrated almost solely on things and abstract relations—machines, equations, physical laws—instead of people. Intensive study of this kind for four or more years tends to drive a man's interests inward, away from people. Yet in the world of technology and business the professional man must deal with both people and things. So we find that getting along with people is often as important as knowing the correct equation or constant. Some management authorities believe that getting along well with people (good *human relations*) is more important than technical knowledge of a field.

Look at how nonengineers describe engineers in a personality profile developed by Gaynor & Ducas, Inc., an advertising agency:

> They (engineers) tend to be more thing-minded than people-minded. . . . They do not delegate authority well. They are not too adroit in personnel relations, nor tender of "feelings" when they get in the way of completing the job. . . . Engineers *prefer* methods, objects and processes to either ideas or people.[1]

Do you see yourself in this excerpt from this study? If so, training in human relations will be a big help to you. For as Maj. Bronson Foster, while president of Alexander Hamilton Institute, said:

> The management level demands of a man the ability to work as a member of a team. The engineer who goes into management can no longer withdraw into his cocoon and study the problem at hand as if it were in a vacuum. He can no longer be a "loner" as his classroom and early job training has often taught him to be. He cannot study his problem with the time-consuming thoroughness to which he is accustomed, because management demands decisions, and the manager must balance the importance of accuracy of decision against the penalties for indecision and delay." [2]

Every engineer and scientist who wants to advance in either the technical or managerial sphere must become adroit at human relations. For if he doesn't develop this skill, his prospects for growth are severely limited.

Human Relations Defined

There are many definitions for human relations. Some are complex; others simple. Perhaps the simplest is that given by John Perry, Consultant in Human Relations. His definition is as follows:

> . . . the real meaning of good human relations is simple enough; people getting along well together. Good human relations in an industry has a special meaning because a group of people has come together for a purpose. The measure of their human relations is whether they achieve that purpose, or whether their effectiveness is cut down by arguments, misunderstanding, hostility, suspicion, and noncooperation. . . . Good human relations for management means knowing how to work with groups of people, and with individuals.[3]

Human relations is *not* smiling and backslapping. Neither is it control or manipulation. It is realistic, practical—compatible with the aims of business and technology. Professor Keith Davis, Arizona State University, speaking to the American Society of Training Directors said, "The objective of good human relations is not to create one big happy family. Its objective is simply to provide an environment for the pursuit of happiness. Some employees will pursue it and find it. Others may do neither."

L. A. Kilgore, director of engineering, and V. B. Baker, assistant director of engineering at the East Pittsburgh Division of Westinghouse Electric Corporation, organized and presented a course in human relations for engineers and supervisors in their company.[4] Courses these two men were familiar with were not particularly suitable for engineers. So they set about organizing a course of their own, drawing freely on available source material and personal experiences.

Specifically, Kilgore and Baker felt the course should be down-to-earth and should promote class discussion. They also felt it should not delve too deeply into the whys of human behavior but rather should concentrate on specific suggestions for improvements in human relations that every engineer could adapt to his own needs.

A large voluntary turnout and active participation (75 per cent of the engineers in one group) showed a high interest in the subject. This course, and many others in industry and government, show that engineers and scientists *are* interested in human relations and *do* profit from a study of the techniques of getting along with people. Let's see how you can apply many of the Westinghouse course findings in your daily activities. Adroitness in human relations can mean as much as a master's degree in your technical specialty.

People, People, Everywhere

Engineers and scientists are often pictured as men with slide rules or test tubes—men of figures and facts. This is only half the picture. For everywhere the engineer or scientist turns, he must deal with people—other engineers, scientists, salesmen, technicians, his own boss. To do his job well, to win acceptance for himself and his idea, the engineer or scientist must be able to deal effectively with all these people. Regardless of your specialty you must deal with people if you expect to move ahead.

As an engineer today you are, more than ever, a team worker. You are teamed with scientists to work on development problems, with sales and shop people in the planning and production of apparatus. Your supervisor may call on you to work with customers or clients on complex application and operating problems. Few engineers and scientists have had formal education in human relations. Many have concentrated so much on their technical or scientific work that they haven't had time to study the human aspects of technology and business.

In today's highly technical and military world you as an engineer or scientist are in a unique position to provide leadership. You are the man with the facts, the analytical methods, the creative mind. It is natural then that you will move into some type of managerial or supervisory position—either in a technical or business category.

All experienced engineers and scientists must often exercise leadership. Some men are natural leaders. You, too, can develop leadership qualities by becoming more effective in your human relations.

Think—and Learn

During your career you've probably met at least one man with whom most people enjoy working and whose ideas are almost always well received. Stop for a moment and think about this man. Analyze his actions and characteristics.

- Does he smile frequently? Or seldom?
- Is his voice friendly? Or distant?
- Can he remember people's names? Or does he forget?
- Is he fair to everyone? Or only friends?
- Are his convictions firm? Or weak?

The chances are good that you'll answer yes to almost all the questions on the left. For without these traits this man could hardly be a successful leader.

Think some more. This time reflect on your own *emotional* reactions to the acts of other people.

- What made you feel good toward someone?
- Why did you want to cooperate fully with him?
- Who were the people he led to happy careers?
- When did he make you feel like working harder?

Analyze the answers to these questions, and you will see that there are *some* general rules of good practice in human relations.

Are All People the Same?

No; personalities differ greatly. But remember: There are many needs, wants, and reactions common to all people, regardless of their personalities. So we *can* develop certain *general* rules for dealing with people.

Is it dishonest to pattern your behavior after a standard set of rules? Some people—particularly certain engineers—say yes. *These people usually cannot learn to apply the rules of human relations naturally and with complete sincerity.*

For example, offering a forced or undeserved compliment to an associate frequently leaves both of you with a bad taste. Your associate, recognizing that the compliment is undeserved, is suspicious of your motives. You are bothered by your conscience for being dishonest.

But take a deserved compliment, sincerely delivered. The compliment improves you and your associate's attitude. How often we overlook a chance to give a sincere compliment. And all normal people react favorably to a sincere compliment. So we can say all people are the same—in some respects. Knowing this you can develop *general* rules for better human relations. But to be effective you must:

- Develop a genuine interest in other people
- Be sincere and natural in all relations
- Adapt general rules to *your* personality and character

As an engineer or scientist you will have to work steadily at these three guides if you wish to become proficient in human relations. Here's how.

Get interested in people. Develop a *warm, genuine interest* in all the people with whom you deal. People instantly respond to *genuine* interest in themselves, their work, their families, their hobbies. And whenever you become interested in a person you benefit substantially too—not only by acquiring broader interests, but also from the personal satisfaction you gain. When this warm interest in people becomes a habit, you subconsciously apply many effective techniques in dealing with others.

How can you develop more interest in others? There are hundreds of ways. Here are a few.

1. "Collect" degrees—ask every new engineer or scientist you meet the name of the school where he obtained his degree. You'll learn a number of interesting facts about schools, degrees—and people. For school talk can lead to talk about sports, educational methods, business, and a variety of other topics.

2. Play a mental sizing-up game with every new person you meet. Try to guess how many children he has, what his hobbies are, which professional societies he's a member of, etc. Playing this game will help you develop more interest in the man—it will also increase your analytical ability. See Chapter 15 for additional hints on judging people.

3. Try to find something worth complimenting in every person you meet next Monday. If you like a man's hat, tell him so. The same goes for his shoes, suit, work, home, or what have you. But don't compliment unless you mean it. An undeserved compliment isn't worth the breath used to utter it.

4. Take Disraeli's advice: "Talk to a man about himself and he will listen for hours." If you *must* talk business, talk about *his* business. Almost every engineer and scientist is as much interested in his business as he is in himself.

5. Be friendly every waking moment of every day. Push criticism of others out of your mind, unless you're meeting with a man for the specific purpose of criticizing him, his work, or some other aspect of your relations. Keep your mind open, friendly. For as Robert Tate said, "A person's mind, like a parachute, works best when it is open."

Recognize importance in others. A *feeling of importance* is a strong human want in most of us. So if you recognize that the other fellow is important and *make him feel that way*—without resorting to insincere techniques—he may exhibit a receptive attitude toward you. Little things—like remembering his name, giving a sincere compliment, or an expression of confidence in him—help bring him closer. Too often we overlook or forget these important "small" things in daily contacts.

Be enthusiastic. Enthusiasm is contagious and inspires others to greater effort. Too often engineers and scientists hide their enthusiasm behind a cloak of "sophistication." When dealing with other engineers and scientists remember that a critical attitude may be an indication of interest that can be turned into enthusiasm. Charles Schwab, one of America's most successful engineers said, "I consider my ability to arouse enthusiasm among men the greatest asset I possess. The way to develop the best in a man is by appreciation and encouragement."

Maintain a positive attitude. Express ideas, tasks, and other assignments in positive terms. By taking the positive approach you help yourself and others to find the best answer to a problem. Be as analytical and critical of errors as you wish. But once your analysis and criticism are finished, take positive steps to successfully complete the task you are considering.

Give others a challenge. This will often help you inspire and motivate people to their highest efforts. People usually respond with their best effort when they know someone has confidence in them. When faced with a difficult problem, a challenge can stir a man's mind and urge him to extraordinary achievement.

Devise your own techniques. You can develop many other ways to improve human relations. Here are three important techniques for engineers and scientists who tend to be overcritical of other people. (1) Let the other fellow save face. (2) Consult with others when the need arises. (3) Show respect for the ideas of your associates.

No matter how many techniques you develop, keep one thought foremost in your mind: *Human relations is an art—not an exact science.* A coldly scientific analysis of personality is a poor substitute for a warm human interest in the other follow.

Adapt your techniques. As you've read this chapter, you have probably said, "But these techniques won't work for me. My situation is different." Sure your situation is different. But not so different that you can't apply *some* of the techniques. One smart engineer faced with a different situation suggested three approaches for dealing with people: (1) the personal interest approach, (2) the objective, or impersonal approach, and (3) the hardboiled "do it or else" approach.

So far in this chapter we've discussed only the first approach. You can use the second or objective approach in many daily contacts where the other person is already interested and is ready and willing to do what is right. You'll also find this approach useful when there is danger of losing control of your emotions. Try the technique one engineer uses to obtain this objectivity by thinking, "How will this situation look to me a week from now after things have cooled off?"

Engineers and scientists rarely need the hardboiled approach for they usually deal with reasonable people. But in rare cases you may meet a person so ruthless that he will not respect you until he is shown that you will not be stepped on or ignored. But give the other man the benefit of the doubt first. Often we misunderstand another's motives.

Don't use the hardboiled approach when dealing with creative people. Driving people by pressure or fear is poor motivation. Resentment and fear of failure tend to inhibit creative thought and distort judgment.

Remember: Interest in the other follow and consideration of his ideas do not weaken your position nor make you any less forceful. An atmosphere of free interchange of ideas actually helps you to gain cooperation and support. People will respect you for an open mind if (1) by listening to others you learn something, *and admit it*, and (2) *if you admit your mistakes without excuses and alibis.*

You Must Communicate

Never neglect communications. What and how much to communicate up and down the line are worth much thought. But most important to engineers and scientists is *how to communicate.* Engineers and scientists have a problem common to all specialists—that of "talking the other fellow's language." Put yourself in the other man's place by imagining that you are sitting on the *other* side of the desk. Choose those words and concepts that will get your ideas across quickest and with minimum effort on the listener's part. Use this technique whether you're talking to a shop man or manager.

Communication isn't only talking. There is another important phase—listening. This listening phase forms the feedback in communications. Try to sense the response of your listener. Use this response to regulate and guide your presentation. Observe his voice and expression, his words, questions, remarks. This technique puts you in a much better position to control the situation than if you just plunge ahead blindly with your own thoughts and feelings. Sensitivity to others' feelings—a constant awareness of the other man—is a basic human relations skill. Practice this skill every time you deal with others—be these people company presidents, chief engineers, messenger boys, or your own children. For the more you practice, the greater your competency and the better you will be liked.

Develop Effective Presentations

To communicate effectively you must make effective presentations. As an engineer or scientist you are already skillful in logical thinking and organization. But as a typical technically trained person you may not be adroit at verbal expression or efficient salesmanship. Presenting a proposal to your superior or making a request for approval of a project is really an attempt to sell your ideas and yourself. Here's a useful checklist for making more effective presentations:

1. *Get a receptive mood.* Spend a few minutes getting the other fellow in the proper frame of mind to say yes. Do this by stressing the positive aspects of the project or proposal. Speak in terms the other man is interested in—use financial aspects when presenting a proposal to management. Express the proposal in technical terms when talking to top engineers and scientists. Take care to time your proposal correctly. If you see that the boss is busy or worried, delay your proposal until a later date, if possible. When you can't delay the presentation, begin with a statement that shows you realize he's busy or worried. Thus, you might start by saying, "I know you're very busy today, J. G., but this proposal has to be approved by noon. If there was any way of putting it off until next week, I'd certainly do it. . . ."

2. *Get attention for the proposal.* Start in an area of mutual interest, related to the proposal. In general, avoid a discussion of the other man's golf game, your latest hunting trip, etc. These discussions waste valuable time and divert your listener's attention. If you believe that some personal banter will establish better relations, leave it until after the proposal is approved. Remember: The purpose of your presentation is approval of an idea. So concentrate on getting approval. Of course, if your boss wants to engage in some personal chitchat before beginning the business discussion, you'll have to listen and respond. But don't you start the chitchat.

3. *Build interest in the subject.* State the subject clearly, and tell him why he should be interested. Tailor your terms to your man, as in item (2).

4. *Build desire.* Appeal to his wants, his principles, his reason, his emotions. If you can, dramatize your proposal with a human interest story that is to the point. When talking to management personnel stress efficiency, reliability, etc. But use your judgment—don't try to oversell.

5. *Leave a clear understanding of what is to be done.* Get him to state what he will do, summarize it yourself, or give a command—whichever the situation demands.

Keep Human Relations in Mind

Now that you've nearly finished this chapter, you're ready to begin an active human relations improvement program. Reading this chapter will help make you aware of a few of the techniques you can use. *But you won't improve unless you begin now consistently and constantly to practice the art of human relations in every phase of your life.* So begin with the next person you meet. You'll be certain to improve rapidly if you:

- Express and show warm interest in people and their problems
- Be as impartial as you can in all dealings with people
- Treat everyone you meet as an individual
- Show genuine appreciation whenever it is deserved
- Always be firm and fair in dealing with others
- Look for what others *can* do; not for what *you* want

REFERENCES

[1] "An Engineer's Profile," *Product Engineering,* June 16, 1958.

[2] Maj. Bronson Foster, "What it Takes to be an Executive," *Petroleum Refiner,* November, 1957.

[3] John Perry, *Human Relations in Small Industry,* Small Business Administration, Washington, D.C., 1954.

[4] L. A. Kilgore and V. B. Baker, "Human Relations and Engineers," *Westinghouse Engineer,* July, 1957.

HELPFUL READING

Argyris, Chris, *Understanding Organizational Behavior,* The Dorsey Press, Inc., Homewood, Ill., 1960.

Bittel, Lester R., *What Every Supervisor Should Know,* McGraw-Hill Book Company, Inc., New York, 1959.

Davis, Keith, *Human Relations at Work,* 2d ed., McGraw-Hill Book Company, Inc., New York, 1962.

Dubin, Robert, *Human Relations in Administration: The Sociology of Organization,* 2d ed., Prentice-Hall, Inc., Englewood Cliffs, N.J., 1961.

Famularo, Joseph J., *Supervisors in Action,* McGraw-Hill Book Company, Inc., New York, 1961.

Fleishman, E. A., *Studies in Personnel and Industrial Psychology,* The Dorsey Press, Inc., Homewood, Ill., 1961.

Gardner, B. B., and D. G. Moore, *Human Relations in Industry,* 3d ed., Richard D. Irwin, Inc., Homewood, Ill., 1955.

Gilmer, B., *Industrial Psychology,* McGraw-Hill Book Company, Inc., New York, 1961.

Grimshaw, Austin, and J. W. Hennessey, *Organizational Behavior: Cases and Readings,* McGraw-Hill Book Company, Inc., New York, 1960.

Heckmann, I. L., and S. G. Huneryager, *Human Relations in Management,* South-Western Publishing Company, Cincinnati, 1960.

Hepner, H. W., *Perceptive Management and Supervision,* Prentice-Hall, Inc., Englewood Cliffs, N.J., 1961.

Herzberg, Frederick, Bernard Mausner, and Barbara Synderman, *The Motivation to Work,* John Wiley & Sons, Inc., New York, 1959.

Leavitt, H. J., *Managerial Psychology: An Introduction to Individuals, Pairs, and Groups in Organization,* The University of Chicago Press, Chicago, 1958.

Pigors, Paul, and Faith Pigors, *Case Method in Human Relations: The Incident Process,* McGraw-Hill Book Company, Inc., New York, 1961.

Sayles, L. R., *Behavior of Industrial Work Groups: Prediction and Control,* John Wiley & Sons, Inc., New York, 1958.

Schleh, E. C., *Getting Results from People,* Prentice-Hall, Inc., Englewood Cliffs, N.J., 1961.

Spates, T. G., *Human Values Where People Work,* Harper & Brothers, New York, 1960.

17

Develop Your Leadership Abilities

> There are no reluctant leaders. A real leader must really want the job. . . . If you find need for a leader and have to coax or urge your selection, you'll be well advised to pass him over. He's not the man you need.
>
> LT. GENERAL IRA C. EAKER

You must develop your leadership abilities if you expect to achieve more in technology, science, or related fields. Engineers and scientists can profit handsomely from study devoted to developing their leadership skills. In this chapter you will learn many of the techniques of good leadership. Use the hints given and begin a program that will develop your leadership abilities to their fullest.

Need for Training

Why can engineers and scientists benefit from leadership training? Here are some findings that show the traits that need strengthening by the average engineer and scientist. Dr. Anne Roe, while a Harvard psychologist, reported that scientists[1] have high self-centeredness or egocentricity. They are not gregarious, not very talkative and rather asocial. Scientists show much stronger preoccupation with things and ideas than with people.

Dr. Floyd L. Ruch, President of Psychological Services, Inc., states,[2] "An excellent engineer may be a second-rate, or even a third-rate administrator, and the company that promotes him to a managerial job suffers doubly. Exchanging a good engineer for a poor executive results in decreased efficiency and morale as well as loss of scarce talent."

Lawrence A. Appley, while president of the American Management Association said: [3]

> When an engineer becomes a manager two specific and dramatic changes must take place in his thinking: He must think in terms of the direction of the efforts of people rather than the completion of engineering projects; he must realize that his progress as a manager is dependent upon the number of people whom he can successfully motivate rather than the number of projects which he personally can complete satisfactorily and on schedule.

These, and many other studies, show that the engineer or scientist must change his ways of thinking if he is to lead people. Since the techniques of leadership are much the same whether a man specializes in technology or business, knowledge of these skills will help a man regardless of his specialty.

What is Leadership?

W. C. H. Prentice, while dean of Swarthmore College, defined leadership as:

> . . . the accomplishment of a goal through the direction of human assistants. The man who successfully marshals his human collaborators to achieve particular ends is a leader. A great leader is one who can do so day after day, and year after year, in a wide variety of circumstances. . . . His *unique* achievement is a human and social one which stems from his understanding of his fellow workers and the relationship of their individual goals to the group goal that he must carry out. . . . When the leader succeeds, it will be because he has learned two basic lessons: men are complex, and men are different.[4]

This excellent definition of leadership shows why most engineers and scientists must work at revising their normal attitudes if they are to succeed in directing others. Let's see how you can develop desirable leadership traits.

Qualities of Good Leaders

Two Westinghouse Electric Corporation engineers, L. A. Kilgore and V. B. Baker, analyzed leadership among engineers and developed interesting findings: [5]

> Leadership is one phase of human relations in which everyone is interested. There is a good deal more to leadership than just getting along with people, being persuasive, and having good communications—though these are certainly a part. *Integrity* is obviously a most important quality of leadership; but there are other qualities of an effective leader.
>
> *Respect of the group:* People like to be led, but they want their leader to have at least one strong point—such as ability, or diplomacy, or courage. Whatever the quality, it must be something that other people recognize. No engineer or scientist should hesitate to stress his own strong points; however, this requires great amounts of tact, diplomacy, and understanding. Make your good points show, but not to the extent of being overbearing or inconsiderate.
>
> *Willingness to take responsibility:* Again, others are usually glad to let you take the responsibility if you demonstrate the ability and the willingness, without being over-eager or reaching far beyond your ability.
>
> *Ability to create a team:* People *like* to be part of a good team. You as the leader have to demonstrate your interest in the group as a team and let them know what you expect of them as part of your team in accomplishing the objectives of the group. Competition with another group helps build team spirit but it can also work against cooperation. Competing with the accomplishments of another organization in the same field or with your own previous record is excellent.
>
> *Ability to correct faults tactfully:* No one likes criticism, but most people will accept correction, in private, if you display understanding and are fair in your comments.
>
> *Devotion to the cause:* You must make it evident that you believe in your stated objectives; then your example will be a major

factor in motivating others. Enthusiasm is contagious. Perseverance provides stability in time of difficulty.

Intellectual courage: In engineering leadership, you must face all the stubborn facts, weigh the risks of the unknowns, and make a decision. People respect and will follow an engineer who has a purpose and a plan, and the courage to act on his convictions.

Are leaders born? Alfred J. Marrow, President of Harwood Manufacturing Corporation, asked the same question while studying leadership.[6] His answer "We do not find scientific evidence to support 'leadership by natural ordination.'" Leaders can and are being trained in industry and education today.

Check Your Leadership Abilities

Here's a comprehensive checklist of your leadership abilities. Study this checklist carefully and then rate yourself as objectively as you can by answering each question. If some questions do not apply to you, skip them. Try this list now—you will be rewarded many times over by the new look you have at your own personality.*

LEADERSHIP CHECKLIST

	YES	NO	
1.	______	______	Do you firmly believe in the true worth of your work?
2.	______	______	Are you ready to accept responsibility for your work?
3.	______	______	Do you show enthusiasm for your work?
4.	______	______	Can you muster courage to defend your employee's work?
5.	______	______	Do you try to emphasize the best in your personality?
6.	______	______	Do you try to develop the best in the personality of others?
7.	______	______	Do you allow your employees to work on their own?

* *Note:* Where employees are mentioned in the checklist this refers to anyone working for or with you and receiving directions from you. Thus, you should include office boys, secretaries, assistants, draftsmen, designers, and engineers and scientists receiving directions from you. The pattern of the questions is arranged to stimulate thinking and self-examination.

	YES	NO	
8.	_____	_____	Are you a persistent worrier?
9.	_____	_____	Is your approach to your job realistic?
10.	_____	_____	Can you make compromises at work when necessary?
11.	_____	_____	Do you insist on having your own way most of the time?
12.	_____	_____	Are you upset by interruptions at work?
13.	_____	_____	Do you analyze every mistake you make?
14.	_____	_____	Do you conscientiously try to learn from your mistakes?
15.	_____	_____	Are you a good excuse-maker?
16.	_____	_____	Can you hold your temper when things go wrong?
17.	_____	_____	Do you give genuine credit to employees for work well-done?
18.	_____	_____	Can you praise people warmly and sincerely?
19.	_____	_____	Do you adjust easily and quickly?
20.	_____	_____	Do you complain regularly about anything?
21.	_____	_____	Do you complain about your organization in front of employees?
22.	_____	_____	Are you biased against any race, color, or creed?
23.	_____	_____	Do you try to get all the facts before making a decision?
24.	_____	_____	Do you boost the organization whenever you can?
25.	_____	_____	Are you sincerely loyal to the organization?
26.	_____	_____	Can you work well with other departments or divisions?
27.	_____	_____	Do you keep your employees posted on organization and department objectives?
28.	_____	_____	Can you keep confidential information to yourself?
29.	_____	_____	Are you willing to delegate tasks to your employees?
30.	_____	_____	Can you correct employees gently and quietly?
31.	_____	_____	Do you try at all times to discipline fairly?
32.	_____	_____	Do you regularly plan your work in advance?
33.	_____	_____	Are you courteous to everyone at all times?
34.	_____	_____	Do you regularly look for superior ability in your employees?

	YES	NO	
35.	______	______	Do you try to assign employees to tasks equal to their abilities?
36.	______	______	Can you express your ideas and opinions clearly and concisely?
37.	______	______	Do your work habits set a good example for others?
38.	______	______	Can you balance sternness and familiarity?
39.	______	______	Do you admit your mistakes to others, when necessary?
40.	______	______	Do you dress neatly at all times?
41.	______	______	Can you listen to others with sympathy and sincerity?
42.	______	______	Do you train your employees for their work?
43.	______	______	Do you try to treat everyone as an individual?
44.	______	______	Do you genuinely welcome suggestions from your employees?
45.	______	______	Do you show interest in guiding your employees?
46.	______	______	Do you try to keep every promise you make?
47.	______	______	Do you make prompt and fair decisions?
48.	______	______	Are you easy to see when employees have questions?
49.	______	______	Are your instructions clear and concise?
50.	______	______	Do you show that you know and understand your employees?
51.	______	______	Do you act like you know your job?
52.	______	______	Do you encourage creativity in your employees?
53.	______	______	Do you encourage employees to improve themselves?
54.	______	______	Do you assign blame for errors fairly?
55.	______	______	Can you discipline people fairly?
56.	______	______	Do you try to mentally put yourself in the other person's place?
57.	______	______	Do you try to keep learning more about your work?
58.	______	______	Do you try to create a happy atmosphere in your work area?
59.	______	______	Can you take orders without resentment?
60.	______	______	Do you praise your employees in public?

Score yourself on the basis of the number of yes answers: 50–58: You are an excellent leader; 40–50: You are a good leader who can improve with study and practice; 30–40: You are a fair leader but you must spend more time improving your abilities; 20–30: Your leadership skills are just developing—work regularly at improving them; 10–20: You would be wise to attend one or more courses in leadership and human relations; 0–10: Face facts—you can improve but you must begin now—enroll for a formal course in leadership immediately.

Program for Leadership Skills

There is no quick, cheap, and easy way to master the art of leadership, states Russell H. Ewing, of the National Institute of Leadership.[7] But if a person has intelligence, natural aptitude, and native interest in the field, he can learn the principles and techniques of leadership and apply them. Leaders are trained and developed, says Mr. Ewing, who has devoted much study to the art and history of leadership. He has developed fifty keys to successful leadership in his research and direct observation of outstanding leaders in many fields. But he warns that it takes far more than memorizing a few principles of leadership to make you a successful leader.

You must understand the principles, live them, and apply them until they become habits of thought and behavior in your daily experience. Instead of leaning heavily on a few skills in which you are superior, you must try to discover your latent abilities and develop your known skills to a higher degree. Each day cultivate some new technique until you perfect it. Learn one new skill at a time and learn it well. When you take part in group activities of any kind or competitive tasks, apply what you have learned about the art of leadership. Now here are Mr. Ewing's fifty skills you can develop.

1. *Master the simple first.* Much of the failure and frustration in leadership stems from young people who try to start too near the top. If you're still young, you'll find it's much wiser to learn leadership in simple situations, or in small groups. Then train for the more difficult tasks.

2. *Never stop learning how.* You can improve your leadership skills and techniques in several ways. Read books on leadership (see

the bibliography at the end of this chapter). Take college courses and attend institutes of leadership; observe and talk with successful leaders. Seek to gain practical experience in leading others in actual and simulated situations. Each new level of leadership requires more information and skill. These you must acquire. There is no end to what you can learn about leadership.

3. *Lead by training others.* The leader who doesn't fear competition and has the best interests of his organization in mind always trains capable assistants who some day will assume leadership roles. To do this, break down each job into simple operations. Review the instructions and repeat the operations until the trainee has learned how to perform each task properly.

4. *Look for leadership in others.* As our society and industry become more complex, they require more capable leaders to manage their affairs. We can meet this need if engineers, scientists, businessmen, educators, and other leaders will look for gifted young people who can be trained for leadership roles in the nation and community. Young people demonstrating high aptitude for leadership through identification tests and other screening devices are our greatest national asset in industry and the community.

5. *Cultivate the right climate.* Aspire to leadership tasks at work, home, school, church, club, party. Exploit situations that enable you to use the leadership talents and techniques you've acquired. *Let your light shine so others recognize your qualifications for leadership.*

6. *Be faithful to principle.* It's been said before—even smacks of self-righteousness. But the safest path to leadership, you'll find, is through unselfish service to others. Our greatest leaders are humble, devoting themselves fully to their chosen work. Egotism and irresponsibility defeat the intent of progressive leadership.

7. *Be buoyant.* Keep your thinking flexible. Rules, ruts, and routine ruin many leaders. Avoid blind-alley jobs and dead-end careers. Observe the customs of the group, but avoid becoming tied down by precedents. High vision is essential, change is inevitable, and true progress is desirable. The mossback in engineering and the standpatter in science will always block the road to greater achievement.

8. *Be a good follower.* Almost every leader is a follower of some superior. To become a leader, learn to be a loyal follower of your present superior. To learn how to give orders, learn first how to take

orders and execute them. Learn how to coordinate the activities of others by cooperating with those you come in contact. Learn teamwork if you want to become captain of the team. Take part in competitive games and group activities in which you can learn how to follow. You'll become a better leader for it.

9. *Learn to like people.* Affection for man is an important basis for true leadership. It may be enough to like your friends. But you'll also have to learn to understand your enemies if you want to defeat them. To learn how to like everybody requires great patience and perseverance, but it is the only course to follow. As John W. Gardner wrote "We must have respect for both our plumbers and our philosophers or neither our pipes or our theories will hold water."

10. *Be yourself all the time.* A high personality quotient, revealed by tests or direct observation, is most desirable. But if it took no more than that to make a great leader, television and movie stars would be running our nation's business. Significant individuality—or a strong personality, well-integrated—is helpful in many kinds of leadership. Ready wit, a nimble mind, a good memory for names and faces—these are useful qualities for industrial leaders. But don't lean too heavily on personality alone. Strengthen your professional knowledge and technical skill also.

11. *Earn respect.* Position-holders and those who inherit wealth or office are not necessarily leaders. The mere fact that you have been appointed to some position doesn't ensure your success as a leader. But it is an opportunity to prove that you are one. Don't stand heavily on your title, your dignity, or your authority.

12. *Inspire others.* A leader knows how to get things done through people. To get results, find the person who wants to do the job and has the ability to do it. The reluctant follower isn't much help to the leader. A better way is to appeal to your follower's self-respect, arouse his ambition, stimulate his imagination, and encourage his initiative. Wise leaders read the newest and best books on the principles of communication and the methods of motivation. Some authorities say the best way to move people to action is to make them laugh, make them mad, or make them cry.

13. *Be enthusiastic.* Practice positive modes of behavior—a happy outlook, a frequent smile, a warm handshake, emotional stability. Consciously cultivating these traits through training and practice will make you a more interesting person. The stiff-as-a-poker, cool-

as-a-cucumber, aloof-as-the-sphinx individual can't lead well. Anyone who wants to ignite the fires of conscience in the followers must throw off a few sparks. To be exciting, let your enthusiasm overflow and your light shine.

14. *Be confident.* A confident air inspires the respect you need to control others. Be sure of your facts, understand the issues raised, and the principles involved in the situation confronting you. Then you'll feel confident and act the way you feel. This confidence will convey conviction to others. Bluffing, braggadocio, and bombast don't beget confidence.

15. *Match people to the job.* Interest is as vital as intelligence and skill. So try to discover the basic interests of people through interest tests and other devices. Influence those you work with by discovering what they want or what pleases them. Where two plans of equal merit are presented for your consideration, select the plan developed by the person who will be chosen to carry it out. Interest in the plan is the seed you sow to help ensure its successful completion.

16. *Expect the best of people.* If your subordinates and assistants are carefully selected, trained, and supervised, you (their leader) will know what to expect of them, and you'll feel you can place full confidence in their abilities. They'll respond by turning out a first-rate job or a high-grade product. Most people seek to win and deserve the confidence that is reposed in them.

17. *Keep your poise.* Lack of poise is caused by careless mental habits. If you lack poise—some engineers and scientists do—work to deliberately acquire it through psychology, practical experience, and faith. You *can* conquer a sense of fear, failure, and frustration, or doubt, dismay, and defeat through conscientious study and self-analysis. Inner tranquility produces outward calm, grace, and charm. The simplest way to achieve assurance is to think about others instead of about yourself. Conscientiously avoid being a head-scratcher, a fingernail-biter, a table-tapper, or a ceiling-gazer. These mannerisms invariably reveal a lack of poise and self-assurance.

18. *Be active.* The dynamic leader is his own self-starter. He's always open to creative inspiration and intuition; and he translates thought into action. He learns how to dramatize action through salesmanship, showmanship, and sportsmanship. So be active without being aggressive. Start things—but be sure you finish them. Follow through with verve and enthusiasm. You *can* conquer apathy and

lethargy by cultivating habits of mental and physical activity.

19. *Be humble*—but not too humble. Never belittle anyone—openly or secretly. See others as they are in reality. Neither underestimate nor overestimate your own or another's ability or mentality. Don't belittle yourself, your present position, or your future possibilities. Personal vanity, executive conceit, and family pride are mild forms of self-delusion in which no leader can afford to indulge. Be as careful of what you say to yourself as you are about what you say to others.

20. *Be consistent.* Nobody can follow a leader who gyrates wildly. The person who is a firebrand one moment and an iceberg the next, who is cordial one day and crusty the next, or who flatters one week and frightens the next, bewilders his followers. The better way is to be even-tempered, consistent in attitudes, and predictable in actions. These traits attract followers and harmonize human relations.

21. *Be gracious.* Courtesy and kindness pave the way to progress in leadership and pay big dividends in terms of goodwill and loyal support to those who unfailingly practice these rules. The true leader is as courteous to his opponents as he is considerate of his supporters. Bickering, curt, and condescending attitudes ruin the careers of many men who are otherwise qualified for responsible executive positions.

22. *Know your organization.* Some authorities state that leadership is a phase of organization. To be most effective, the leader should know thoroughly the details of his organization—its policies, procedures, personnel, methods, and functions. This rule holds true in all phases of your life—professional, business, civic, political, religious. So be sure you know your organization, and what your group stands for. Give your people goals—a sense of direction.

23. *Be an attentive listener.* Your superiors, equals, and subordinates deserve the keenest interest and closest attention during personal conversation and group discussion. The adroit leader asks for opinions and pays careful attention to what is said. If you ask for opinions and do not evaluate what is said you learn nothing. People who do this are merely waiting to tell others what to do, regardless of the other person's opinion.

24. *Follow the chain of command.* The engineer or scientist who ignores his superior and takes his ideas directly to the president usually defeats his own purpose by attempts at slicing off his su-

perior's prestige or authority. In making suggestions or giving orders it's wiser to follow the lines of authority downward and the lines of responsibility upward. As was said centuries ago: "No man can serve two masters."

25. *Learn from others.* None of us is an all-around genius, expert at everything. Recognize today that every specialist has his limitations. Be a great leader—learn how to use experts and specialists in many fields.

26. *Be cooperative in seeking answers.* Find acceptable solutions or workable answers to issues, questions, or problems by working closely with others. Remember—leadership can't exist without people. So don't try to be a leader in a social vacuum—it won't work.

27. *Be interested at all times.* Develop a sincere interest in the welfare of others and a warm-hearted appreciation of what they stand for and are doing for the organization. Interest in others is of utmost importance to successful leadership. Show your interest by using first names, inquiring about family news, mentioning hobbies, and other approaches. Show respect for others by avoiding gossip. Don't discuss race, religion, politics, and other topics that are strictly personal. Being genuinely interested at all times strengthens your leadership abilities and position.

28. *Don't show off your authority.* Assume that others are working *with,* and *for* the organization. Ask people to help you. Make suggestions, but issue few orders. Use commands only when persuasion fails. Attitudes of dictation and domination breed lackeys—not leaders. Exercise influence through regulations, training conferences, staff meetings, and personal interviews in which you can display your technical knowledge and professional skill, and seldom through a direct show of authority.

29. *Be thoughtful.* Plan in advance when you're conducting personal interviews or group meetings. Know what you'll say and how you'll say it. Choose your words with care; watch the inflection of your voice; be sure you've made your meaning clear. Many people use an interview checklist to be sure they haven't forgotten a vital point.

30. *Criticize constructively.* Smart people take a hint when you criticize constructively—by correcting errors, reviewing regulations, issuing new instructions, or re-training employees or followers who are making mistakes. Be sure you know all the facts, see all the

angles of the situation, and know how performance or production can be improved by correct practice. Then criticize, if you must. *Where possible, criticize methods and techniques instead of persons.* If personal criticism seems necessary, soften it with a little praise.

31. *Delegate authority.* The wise leader selects and supervises subordinates, delegates to them the details of management, and then evaluates the results. Learn now how to (*a*) delegate authority, and (*b*) hold subordinates accountable for results. You'll be a better leader when you develop these skills.

32. *Admit your mistakes.* A wise leader honestly admits his mistakes and corrects them in humility. He never gives alibis—or blames others for failures. Many great leaders have assumed responsibility for the mistakes of subordinates. If you assume an air of omniscience, you lose esteem when you're found to be fallible.

33. *Be firm but fair to everyone.* The deft leader avoids public criticism of his followers. He uses threats and rebukes only as a last resort to bring people into line. Avoid slighting, scolding, or ignoring others—it isn't good leadership. You can usually avoid the need for any kind of disciplinary measures through proper training, supervision, and the use of modern grievance procedures. Lead and coach; don't club and coax.

34. *Plan programs in advance.* Planning—a blueprint for action—is essential for short-term and long-range leadership. Time your plans carefully—good timing is necessary for effective planning. If you're a conference, committee, or group leader, plan your programs, schedule meetings, and select participants well in advance. As a leader it's your duty to select subjects for discussion, assemble materials, develop outlines, and prepare reports to help the group arrive at decisions and carry out directions.

35. *Study the great leaders.* Study the literature on leadership (see bibliography at the end of this chapter) to determine the specific qualifications and techniques necessary for proper performance of the leadership role you either now occupy or aspire to in the future. Reading biographies and autobiographies of leaders in your field will help you learn valuable techniques from their experiences and avoid costly errors.

36. *Be affirmative and creative.* The greatest leaders are positive, affirmative, constructive, creative, and cooperative in their outlook on life. Avoid being negative, opinionated, abusive, and destructive

in your attitudes toward others. Chronic complaining, grumbling, and criticizing are not conducive to leadership. Avoid these habits at all costs.

37. *Give credit.* Turn the spotlight on the person responsible for special achievements. Never try to take undeserved credit for another's work by hanging the halo of achievement on yourself. People will lose respect for you. By not giving credit to your employees you deaden initiative, dull ambition, and discourage group morale. Give proper recognition for what others have done—if you wish to promote the success of a project. You will collect double dividends—as a leader and from your followers.

38. *Praise your people publicly.* Normal people thrive on appreciation and praise. (Think how you glowed the last time you received a sincere compliment.) To praise privately smacks of flattery. Proper recognition in a public meeting usually inspires people to achieve greater effectiveness. Judicious praise by the appropriate official in a letter of commendation or by award of some fitting emblem boosts the recipient's morale and raises the prestige of the group.

39. *Reprove tactfully.* People, the best of them, make errors or break rules. Reprove privately and informally—it's always better. Reprimands in front of others cause resentment and defeat their real purpose. Avoid hurting the feelings of others, if possible, when taking disciplinary measures—especially where innocent mistakes were made.

40. *Rate fairly—find a yardstick.* Use the 50 hints given in this chapter as a self-rating scale to help yourself to face up to your own performance as a leader. Find a fair yardstick or rating rule to rate others. Apply the rule regularly, remembering that individuals differ, and rules can be broken, when necessary.

41. *Keep people informed.* Don't try to put things over on people or keep them in the dark about matters of vital interest to them. If you expect teamwork, treat people as members of the team, entitled to know what's going on. This builds confidence, avoids suspicion, and strengthens morale by cultivating a sense of belonging.

42. *Respect the work of others.* Show every individual how he fits into the general plan. Get him to see that his well-done work is part of what makes a success of the organization. If he believes he's a mere cog in a machine he may feel frustrated and lose his ambition.

For this reason you'll often need to use orientation, indoctrination, and morale-building programs to help each member see the importance of his job, however modest it may be. We all need to see our own place in the over-all scheme before we start clicking.

43. *Give reasons.* Explain the whys and wherefores of policies and decisions. This way, you'll avoid bickering and backbiting. People want to know why. They want to know what they're doing, how to do it the best way, and why it's the correct way to do it. So wherever possible, give the reasons for everything. People will risk their lives and fortunes on a righteous cause they understand.

44. *Speak directly, briskly.* Be clear in thought and direct in manner without being blunt in speech. Look directly at people while you speak to them. Stick to the subject. Draw out the other person to learn what he's thinking. Otherwise you'll have difficulty in influencing his action. Be brisk without being brusque. Briskness in speech, glances, gestures, and responses can be cultivated and made habitual modes of thought and action through regular practice. Incisive thinking promotes decisive action.

45. *Compromise, but don't appease.* The radical and reactionary are usually in a rut. The best leaders occupy the middle ground between those who lag behind and those who rush ahead of the group. Learn to compose differences and make concessions when this doesn't endanger the ultimate success of the cause and involves no departure from principle. *Don't make a fighting issue of trivial matters.* Preserve your neutrality between small cliques in the organization by helping them to see the common goal. In making compromises you'll be safer in pleasing others than in pleasing yourself.

46. *Be fair to yourself.* Avoid any suggestion of self-punishment, or the temptation to injure yourself with overwork in order to succeed. Rid yourself of silly beliefs that you are unworthy, unkind, or unable to succeed. Such suggestions have only a self-defeating purpose. Root them out of your consciousness. Failure is as often caused by imaginary deficiencies as it is by real ones. Avoid these faults by spending as much time improving your leadership techniques as you previously spent worrying about your shortcomings. Be fair to yourself by helping your own career and the careers of your associates.

47. *Be brave—not brash.* Cowardice in a leader never inspires courageous followers. Overcome fear and cowardice by developing courage in the group. Take courage together. Cowardice and fear

arises mostly from a sense of inferiority, which you can overcome through study and experience. A simple way of overcoming cowardice is to face it squarely—first with confident thoughts, then with brave words, and finally with brave deeds. Learn to do what you seem afraid to do. Fear is faith in evil—don't let a misplaced faith hobble your life and career.

48. *Develop a sense of humor.* Learn to be diplomatic by cultivating tact and a sense of humor. There are few born diplomats. Tact, or a sense of proportion, is the ability to be diplomatic in difficult situations. An objective attitude prevents you from becoming ruffled at rebuffs. It's a priceless leadership trait that you can develop. Learn to display tact by a pleasant word, a friendly glance, or a show of patience. A sense of humor is the ability to see things out of proportion. It's best expressed in ready repartee, a funny story, or a hearty laugh at jokes on yourself or others. It relieves strain, lightens heavy burdens, breaks down barriers, softens antagonisms, and helps you, as a leader, to glide over rough spots and avoid awkward breaks in personal relations. Although these two attitudes view life from different angles, they travel well together and always attract followers.

49. *Be dynamic.* Industry demands dynamic leaders who have the driving energy and determination to excel. Leaders with this dynamic ability must be discovered if industry is to survive in this competitive world.

50. *Cultivate a strong moral sense.* Study of the world's great leaders shows that those loyal to principle seldom lost the support of the people they served. In like manner, our own ability to understand, and our willingness to follow the established rules of leadership without hesitation or reservation, will determine in large measure our capacity to lead others.

Leadership Is Your Key to Achievement

Study the above fifty hints carefully—you cannot expect to be a successful leader without applying many of these hints. And leadership is one of your main keys to professional achievement. Jerome Taishoff, while president of Mycalex Corporation of America, a leading electronics and materials company, advised: [8] "Hire people who know what they are doing . . . then let them do it. My main function (as president) is to set policy and keep the company healthy, not to tell a trained specialist how to do his work. People sometimes need guidance but they shouldn't be pushed."

Taishoff is a strong believer in allowing workers with versatility the leeway of using more than one set of skills and urges his management to give personnel a chance to do more than one kind of job.

Begin developing your leadership skills today.[9] You will find the task as interesting and as rewarding as any technical or scientific assignment you've ever had. If you don't like long rules, then memorize the following short ones: Don't be a watchdog; lead—don't force; never hurt a man's pride; explain everything; set the example; weigh what you say; promote friendly competition; listen well; don't play favorites; stop trouble before it starts; don't gripe; learn patience; be fair; praise good work; avoid prejudice; build the man's job; accentuate the future; avoid destructive criticism; show appreciation; discuss job problems; stress individual initiative; don't make false promises; be consistent.

REFERENCES

[1] Robert C. Toth, "How Scientists Differ from the Rest of Us," *New York Herald Tribune,* Dec. 20, 1960.

[2] Floyd L. Ruch, "Equation: Good Engineer Equals Bad Executive," *Product Engineering,* May 12, 1958.

[3] "Points of View," *Product Engineering,* March 10, 1958.

[4] W. C. H. Prentice, "Understanding Leadership," *Harvard Business Review,* September–October, 1961.

[5] L. A. Kilgore and V. B. Baker, "Human Relations and Engineers," *Westinghouse Engineer,* July, 1957.

[6] Marrow, A. J., *Making Management Human,* McGraw-Hill Book Company, Inc., New York, 1957.

[7] Russell H. Ewing, "Fifty Tips for Better Leadership," *Factory Management and Maintenance,* October, 1957.

[8] People and Plants, "Taishoff: Guide, Don't Push," *Electronics,* Sept. 1, 1961.

[9] "These Get the Job Done for Me," *Petroleum Refiner,* vol. 35, no. 5.

HELPFUL READING

Bass, B. M., *Leadership, Psychology, and Organizational Behavior,* Harper & Brothers, New York, 1960.

Bellows, Roger, *Creative Leadership,* Prentice-Hall, Inc., Englewood Cliffs, N.J., 1959.

Bender, J. F., *The Technique of Executive Leadership,* McGraw-Hill Book Company, Inc., New York, 1950.

Jennings, E. E., *An Anatomy of Leadership: Princes, Heroes, and Supermen,* Harper & Brothers, New York, 1960.

Laird, D. A., and E. C. Laird, *The Technique of Building Personal Leadership,* McGraw-Hill Book Company, Inc., New York, 1944.

———, *The Technique of Handling People,* McGraw-Hill Book Company, Inc., New York, 1954.

Mason, J. G., *How to Be a More Creative Executive,* McGraw-Hill Book Company, Inc., New York, 1960.

Maynard, H. B. (ed.), *Top Management Handbook,* McGraw-Hill Book Company, Inc., New York, 1960.

Odiorne, George, *How Managers Make Things Happen,* Prentice-Hall, Inc., Englewood Cliffs, N.J., 1961.

Petrullo, Luigi, and B. M. Bass, *Leadership and Interpersonal Behavior,* Holt, Rinehart, and Winston, Inc., New York, 1961.

Schell, E. H., *The Technique of Executive Control,* 8th ed., McGraw-Hill Book Company, Inc., New York, 1957.

Schleh, E. C., *Getting Results from People,* Prentice-Hall, Inc., Englewood Cliffs, N.J., 1961.

Tead, Ordway, *The Art of Administration,* McGraw-Hill Book Company, Inc., New York, 1951.

———, *The Art of Leadership,* McGraw-Hill Book Company, Inc., New York, 1935.

Uris, Auren, *The Efficient Executive,* McGraw-Hill Book Company, Inc., New York, 1957.

———, *How to Be a Successful Leader,* McGraw-Hill Book Company, Inc., New York, 1953.

18

Decision Making Can Be Easier

> Every act of *every* human being in an organization can be considered the result of a *decision*—of an operation, either conscious or reflex, of an individual's mind.
>
> HAROLD F. SMIDDY

Successful engineers and scientists have one trait in common—they know how to make intelligent decisions. For no matter what their specialty, interests, or responsibilities, they must be able to make accurate decisions if they are to achieve more in their profession. The same is true, of course, of yourself. You, too, must learn the art of making sound decisions. Without this skill your chances for outstanding professional achievement are extremely low.

Technical Men and Decisions

Most engineers and scientists, because of their training and mental inclination, prefer the neat problem—that is a problem that can be completely defined and solved using known techniques or projections of these techniques. This outlook is useful as long as a man is doing routine work that does not require supervision of others or control of large programs. But as soon as you begin to move ahead in your profession and achieve more, you must develop the ability to make sound decisions. There's just one trouble with the decisions you must make as you move ahead—your past experience has not, in general, prepared you for these decisions. So they are difficult—and the further you advance, the more difficult the decisions become.

Clarence B. Randall, former head of Inland Steel Company, said "It is the man who, by education or broad reading, has opened his mind to the widest range of intellectual experience who has the best chance of recognizing unfamiliar problems before stepping into them." [1]

As an engineer or scientist you *can* improve your decision-making ability. This chapter shows you workable procedures. Apply them as often as possible and you will soon develop better ability to handle the important decisions you meet in your daily activities.

Eight Steps in Making a Decision

Every successful engineer or scientist has a personalized procedure for making a decision. He generally develops this procedure over a period of years. During the developmental stage he will probably make a number of errors. But he is alert enough to profit from these mistakes, thereby avoiding them in the future.

Important decisions often involve similar elements. Thus, many of your difficult decisions will involve people—whether to promote a man or pass him over; fire a man or keep him on the payroll long enough to prove himself. Money, time, and drastic changes in products or procedures will be other factors you must consider. Company policies, taxes, labor relations, sales, community relations, and future plans may also figure in your decisions. From this short list you can readily see that high-level decisions are seldom simple—they can often involve many more factors than the routine engineering or scientific problem. Here are eight useful steps to follow when making a decision. These are based on the latest studies of effective decision making.

1. *Define the problem.* Write it out in as few words as possible. Be as accurate as your information permits. Don't state your problem in such a manner that only one decision is possible. Thus, if the problem is *how to expand manufacturing facilities*, don't define it as:

Should a new plant be built to expand manufacturing facilities. There may be many other ways to expand facilities that do not require building a new plant. Or suppose your problem is one of acquiring a specific amount of capital to buy some new equipment. You might define this as: *How to raise $25,000 to purchase two new punch presses.* This statement mixes money and machines—a concoction that can be misleading. A more generalized definition of this problem is: *How to raise $25,000 working capital.* With this statement of the problem you can concentrate on your key difficulty—finding a specific sum of money. Concentrate on defining your problem in general terms, and you will improve the first step of every decision you make.

2. *Collect data useful in solving the problem.* Seek information from specialists inside and outside your organization. Check past procedures and results with similar problems. Aim your data selection at the critical aspect of the problem. Thus, in the two problems in item (1) the critical aspects are *expansion of manufacturing facilities* and *methods for raising capital.*

When collecting data be careful not to choose facts that will lead you to only one possible decision. Thus, if you limit the facts in the facilities expansion problem solely to those relating to acquisition of a new plant you may overlook important alternatives like purchase of semifinished parts, subcontracting, etc. Or, if in raising capital, you limit your facts to commercial loans, you may rule out chattel mortgages, bonds, stocks, and other sources of capital.

3. *Study all facts relevant to the problem.* Develop as many alternative solutions as possible. *Don't rush—take your time.* Quick decisions are often bad decisions. The more time you spend analyzing data relevant to a problem, the clearer the critical aspects become. Also, you will accumulate more facts as time passes. Biding your time has another advantage—it allows you to work off any emotions you may have about a particular alternative. As time passes, you will be able to see each alternative from a more objective standpoint.

Of course, you will often be pressured for a quick decision. Try to delay important decisions long enough to make an intelligent choice between the alternatives available to you. As Harold F. Smiddy, distinguished contributor to scientific management wrote: "Deciding . . . implies *freedom*—freedom to choose from among alternatives without externally imposed coercion, and freedom to conceive alternatives from which to choose."[2]

4. *List your alternatives; compare the feasibility of each alternative.* Make a dry run on each alternative, i.e., work it out mentally and on paper. To do this, assume you have decided to choose a particular alternative. Write down the results of this choice. List items like capital required, time that must elapse before the objective of your efforts is achieved, manpower needed, etc. Do this for each alternative. Use your engineering or scientific training to tabulate or plot the quantitative aspects of each alternative. Such a comparison will enable you to see more clearly many factors that will be important in your final decision.

5. *Make your decision; choose the best alternative.* Do not announce your decision to anyone. Take time to allow your choice to develop in your mind. Don't be an "If I'd but known" decision maker. Make more dry runs on your choice; follow it step by step to its outcome. Try to foresee problems that may develop along the way. Solve each problem, either mentally or on paper. Above all, bide your time. Some smaller problems may resolve themselves while you study your decision.

Where your decision leans heavily on facts—like costs, production rates, schedules, etc.—review every fact. If assistants supplied facts and data, have them recheck each item. "This takes time; I trust my assistants," you say. Certainly it takes time—every big decision is time-consuming. But the time spent checking facts and data will be worthwhile because you'll (*a*) have a greater chance to study your decision, and (*b*) be more confident of the accuracy of your facts and data. Having your assistants review the facts may reveal errors in their original information or uncover new facts. While you may trust your assistants, never announce a major decision until *after* you have personally rechecked the facts or have had a reliable assistant do so. If you make a major error because your facts were incorrect, your professional career may be seriously harmed.

6. *Choose the best timing for announcing your decision.* Study your organization and decide when the announcement of your decision would get the greatest attention. Thus, mid-July might be a poor time in an organization having a heavy summer vacation schedule. You'll usually find that it is wise to announce your decision at a

meeting attended by the people concerned. Immediately after the meeting a memo detailing the decision should be circulated to the people in your organization affected by the decision.

In government organizations, research laboratories, and similar groups your decision might be best announced on the first of the month. Major reorganizations and similar decisions are sometimes delayed until July 1 or January 1 so a new period can begin in accordance with previously established budgets or other plans. Your announcement, verbal, or written, or both, will precede this date by weeks or days, depending upon what you decide or the particular rules that may exist in your organization.

Lastly, there are times when an immediate announcement of a decision may be most effective. Thus, with a radical shift in policy or procedures your decision might gain more attention when announced and made effective immediately. Or you may be working in an activity—such as the space program—where an hour's delay cannot be tolerated. Under these conditions you would, of course, announce your decision immediately after it was made.

7. *Take immediate action on your decision on the effective date.* Go all out to build full cooperation and maximum enthusiasm for the objective of your decision. Don't allow your decision to die because of inaction on your part. Good decisions which aren't put into action are next to worthless.

You'll find that people will resist decisions which change their work habits. This is a natural reaction for which you must be prepared. So push ahead toward attainment of your objective. True, some feelings may be hurt, but you can soften the blow by using good judgment and exercising understanding and consideration. See Chapter 16 for helpful hints.

Use your technical knowledge of the principle of momentum to keep your decisions moving toward success. Thus, the greater initial enthusiasm you build and maintain for your project, the greater its chances of achieving a favorable outcome. So take action early, and maintain a sustained push until the objective of your decision is reached.

8. *Review your results.* Set up checkpoints along the way so you can easily determine your progress. Even with the best preparation you will sometimes find it necessary to alter your plans. The best way to discover this need is by making regular checks on your

progress. Does your decision have a dollar limitation? Then check your finances regularly. Does it have a time schedule? Then check your progress daily.

Regular review of results will show how and where you stand. Each success on the way toward your objective will spur you on. Your decision will grow in importance, and your desire to achieve your goal will become stronger. With a positive attitude you have far greater chances of accomplishing your task and verifying the accuracy of your decision.

Guides for Sound Decisions

Cities Service Oil Company suggests [3] the following six basic rules for sound decision making:

1. *Avoid impulsive decisions*—accept a solution only when you've arrived at it by a step-by-step process.
2. *Eliminate emotional influences*—such as personal desire, prejudice, inclination, as much as you can.
3. *Scrutinize any decision*—which is too much in line with your own wishes. Life just isn't that easy.
4. *Challenge your first solution*—recheck all facts that support or fail to support it.
5. *Keep your mind relaxed*—the feeling of pressure may tempt you to take short cuts.
6. *Follow the facts*—to the ultimate conclusion.

Use Effective Decision Making Every Day

Good decisions will help you most in your professional and business activities. But don't overlook the importance of decision making in your personal life. Thus, correct decisions can help prevent blunders in the selection and purchase of your home, automobile, and other major possessions. Making a series of wise decisions in your personal life can mean the difference between burdensome financial responsibilities and a manageable responsibility.

So make a regular practice of careful decision making. Apply the principles outlined in this chapter to as many formal decisions as possible. With practice your skills will grow. Then you'll be better prepared for the major decisions when they arise.

Quick Decisions Can Pay Off

Some major decisions you make may appear to be quick, i.e., made with little conscious forethought. But if you analyze these decisions, you will often find that you have been preparing yourself for them for many weeks or months. Thus, you may find yourself in a rut, as Donald Doughty, a University of California mechanical engineering Ph.D. did.[4] Suddenly presented with the idea of building correctly engineered home swimming pools, Doughty grasped the opportunity. Today he is owner of a highly successful pool-construction firm. Though a decision to make a radical change in an important segment of your life may seem to warrant major study, forces at work in your life may subconsciously lead you to a decision. Thus, Doughty left a position with an oil company to found his own firm.

So sharpen your senses; become more alert to the ebb and flow of your goals, interests, and desires. Then you'll be better prepared to make the quick decision when it arises. The chances are, though, that your decision will be well-thought-through. And if you apply the eight steps outlined in this chapter almost every quick decision will be a correct one.

Lastly, if decisions seem difficult to make, remember the words of B. A. Whitney which appeared in *Adult Leadership:*

> There is nothing more human than a reluctance to make important decisions. There is nothing more destructive to human progress than indecision . . . Wrong decisions generally set in motion corrective action . . . Failure to make timely decisions leads only to stalemate and frustration. Strong character and strong leadership come from those who have self-respect, self-discipline, and the courage to make decisions.

REFERENCES

[1] Clarence B. Randall, "The Lonely Art of Decision-Making," *Dun's Review and Modern Industry,* June, 1959.

[2] H. B. Maynard (ed.), *Top Management Handbook,* McGraw-Hill Book Company, Inc., New York, 1960.

[3] *Business Management,* August, 1961.

[4] "People on the Way Up," *The Saturday Evening Post,* July 28–August 4, 1962.

HELPFUL READING

Chernoff, Herman, and L. E. Moses, *Elementary Decision Theory,* John Wiley & Sons, Inc., New York, 1959.

Cooper, Joseph D., *The Art of Decision-Making,* Doubleday and Company, Inc., Garden City, N.Y., 1962.

Greene, J. R., and R. L. Sisson, *Dynamic Management Decision Games,* John Wiley & Sons, Inc., New York, 1959.

Luce, R. D., and Howard Raiffa, *Games and Decisions: Introduction and Critical Survey,* John Wiley & Sons, Inc., New York, 1957.

Mason, J. G., *How to Be a More Creative Executive,* McGraw-Hill Book Company, Inc., New York, 1960.

Melman, Seymour, *Decision-Making and Productivity,* John Wiley & Sons, Inc., New York, 1958.

Miles, Lawrence D., *Techniques of Value Analysis and Engineering,* McGraw-Hill Book Company, Inc., New York, 1961.

19

Delegate Authority to Increase Your Efficiency

It marks a big step in a man's development when he comes to realize that other men can be called in to help him do a better job than he can do alone.

ANDREW CARNEGIE

Talk to any successful engineer or scientist, and he'll tell you that with every promotion comes greater responsibility and a heavier work load. How, you ask, do top men achieve so much when their responsibilities and work load are so great? There is one key answer to this question. *Top men accomplish more by delegating authority to their subordinates.*

Delegation may appear a simple task, but engineers and scientists often have extreme difficulty in "letting go," i.e., in assigning *duties, responsibility,* and *authority* to others. Many engineers and scientists say, "I can't trust anyone but myself to do this work." So long as they maintain this attitude, their chances of accomplishing much are limited. For that reason this chapter deserves careful study by every engineer or scientist who wants to achieve more in his profession. As D. Robert Yarnall, Jr., an outstanding engineer, wrote: "The boss who is most effective in getting best results through motivating his subordinates to high performance is likely to . . . try to provide each subordinate with the resources and authority (which might better be called freedom to act) which are required to get the results for which he is responsible."[1]

Why Delegate?

Here are five excellent reasons why wise use of delegation—the assignment of duties, responsibility, and authority to someone else[2]—will help you achieve more:

> (1) Delegation of routine work allows you to concentrate on the important jobs requiring your skill, judgement, and experience. (2) You then have more time to plan future work, evaluate current problems, and make decisions. (3) Delegating authority helps you develop reliable subordinates who will be better equipped for promotion to positions of greater responsibility. (4) You can then draw on the skills of each of your subordinates. This permits you, and each member of your group, to make a better showing in your organization. (5) Delegating authority gives you more time to control and direct the output of your group, putting you in a stronger position to define and achieve the goals you have set.[3]

What Is Delegation?

Delegation is the assignment of duties, responsibility, and authority to someone else. *Duties*[2] define the task a person is to do. *Responsibility* is his obligation to perform the task and to account to higher authority for his performance. *Authority* gives the person power to act officially within the scope of his delegation.

To delegate intelligently you must analyze your own responsibilities. For if you do not know what your own duties are, you can hardly expect to assign tasks to others.

Before You Delegate

Studies show that many people in industry, government, and other organizations produce at less than 40 per cent of their potential.[3] This condition, if it exists in your group, can seriously hamper your efforts to achieve the goals you've set. You can help people to increase their productivity and enable them to accomplish more in their careers if you know when to delegate, to whom to delegate, and how to delegate. But before you delegate you must understand yourself and others.

What about yourself and others? Do you sincerely believe that: (1) You can achieve more by harmoniously working with and through other people? Much experience shows that you must, with rare exception, work closely with people in your group, if you expect to make more than a routine contribution to your field. (2) You must, when delegating duties, give your subordinates freedom to act. But when you give this freedom, you must still retain control of the results. Without control you are no longer in command of the assignment.

(3) You must learn to identify with your subordinate for long enough to understand his problems when you delegate duties. Seeing and understanding his viewpoint will enable you to delegate more intelligently. (4) You must be (a) polite, (b) interested, (c) confident, and (d) willing to take the blame, when you deserve it, in all situations where you delegate duties. Without these attitudes you cannot obtain maximum cooperation from your subordinates.

(5) You should, at all times, observe a key rule of good human relations: *If you must criticize, do so only in private.* You gain nothing by chewing a man out in public. He will resent you for years and may be a continual roadblock to your achievements. (6) You must give credit freely, honestly. Everyone gains when you take the time and effort to give sincere praise for a job well-done. (7) You *can* arouse a strong desire in your subordinates to improve their efficiency and output. With this desire will come greater enthusiasm and willingness to work for the goals set for your group.

As an engineer or scientist you may believe that a better knowledge of yourself and others is unnecessary—that what counts is the right equation for a specific technical problem or a suitable experiment to prove a particular theory. You must learn now, if you wish to achieve more through others, that knowledge of yourself and others is as important as any technical or scientific aspect of a problem. You must develop the confidence to trust others, many of whom may have as much, or more, ability than you do. Willingness to work with others enables you to bring *many* minds to focus on a problem instead of just your own mind. The success of engineering and scientific teams in recent years reflects the use of many minds to solve an important problem.

When Should You Delegate?

There are specific times when it is wise to delegate work to subordinates. Thus, you will often delegate when:

> (a) a new employee is added to your staff; (b) an experienced employee leaves your staff for more than a few days; (c) you are assigned new duties because you were promoted, transferred, or otherwise reassigned; (d) important actions must be taken to accomplish specific ends in your organization; or (e) a new division, department, or group is formed to perform certain tasks.[3]

Of course, there are other situations in which you will delegate authority. The important thought to keep in mind is that you should be ready to delegate when the circumstances are right. Many studies of executive failures show that a prime cause of poor performance is an unwillingness to delegate authority. As a general rule successful leaders in all fields delegate more than unsuccessful leaders do. But just as there is a right time to delegate, so too is there a wrong time.

When Not to Delegate

Do not, in general, delegate when:

> (a) someone is to be disciplined—do this yourself; (b) important policy matters affecting the long-term performance of your organization are to be decided—these decisions require a mature and experienced approach; (c) high-level decisions of any type are to be made; and (d) new positions are to be created at or about your rank in the organization.[3]

Note an important fact about delegating—you always delegate *downwards* to subordinates. You *never* delegate upwards to your superiors. There are situations in which you might *ask* your superior to do you a favor—for example, give a message to the president of the organization when he sees him at lunch. Your superior has the right to agree or disagree with your request. But you cannot delegate him to carry the message to the president as you might delegate one of your subordinates. So be extremely careful not to attempt to delegate upwards—it doesn't pay.

What to Delegate

You now know when to and when not to delegate. Your next problem is *what* to delegate. Here are a few typical assignments you should consider delegating to talented subordinates: (*a*) routine decisions of the nonpolicy type, i.e., approval of purchases of supplies up to a certain dollar value, choice and execution of designs of specific equipment—for example, motors, pumps, switches, etc.; (*b*) studies, analyses, and tests which you do not have time to perform; (*c*) preparation of routine or special reports you will check and approve before forwarding to a higher level of management; (*d*) hiring of assistants or helpers who will work directly for your subordinate; and (*e*) preparation of budgets, personnel requirements, experiments, and similar items which you will inspect before a final decision is made.

The exact duties, responsibilities, and authority you can safely delegate vary with the ability of the subordinate, the rules of your organization, and your personal beliefs. As a general rule you should begin by delegating simple tasks. When the subordinate completes these tasks successfully, assign new, somewhat more difficult tasks. Continue assigning new tasks until you believe your subordinate is working at maximum efficiency. Thus, you might begin by delegating recurring minor decisions. From these you would advance to more important decisions affecting a number of people in your group. As the final step you might delegate all your responsibilities to a talented subordinate for short periods—say a week or two while you are away for any reason. You must, of course, have complete confidence in your subordinate before delegating him this much responsibility. Always be sure that the subordinate understands that such responsibility is his *only* while you're gone and that when you return, he will relinquish the authority temporarily assigned him.

Choose Delegates Carefully

"To whom should I delegate this" is a constant question facing a manager—particularly if he has a large staff working for him. As you've probably observed, there are subordinates to whom you can delegate a variety of duties which they will perform quickly and efficiently. Other subordinates have limited abilities—they may handle only one or two types of tasks with sufficient skill. So in choosing a delegate for any task you must use extreme care if you want the best results.

What should you look for in subordinates to whom you'll delegate duties, responsibility, and authority? Here are four traits worth watching: (1) Ability—the greater a man's ability the wider the range of duties you can successfully delegate to him. (2) Interest—an interested man with moderate ability may accomplish far more than a disinterested man having outstanding abilities. (3) Willingness—a willing man, again, is many times more effective than the man who must be constantly pushed. (4) Sincerity—the man who believes in what he's doing will produce more effectively than the man who couldn't care less.

So check these four traits the next time you have an important duty to delegate. Know your subordinates. For in knowing them, you will be better able to choose the right man for a specific job. Remember: In choosing the right delegate you achieve six objectives—the job is done; it is done correctly; the delegate acquires more skill and experience; his confidence in himself is improved; you are praised for choosing the right man for the job; an important goal of your group is achieved. As Andrew Carnegie, one of the most successful managers of all time said, "The secret of success is not in doing your own work, but in recognizing the right men to do it."

How to Delegate

Here are ten steps many successful engineers and scientists follow when delegating. Study these steps now and begin applying them immediately.

1. *Get the facts.* Know your goals; have a plan and a purpose. You cannot delegate successfully unless you are familiar with all the important facts. Even knowing that you *don't know enough* to make a decision is important. For if you know this much, you can delegate someone to gather the facts you need.

2. *Decide exactly what you will delegate.* Put your decision in writing. Thus, you might decide to delegate the *responsibility for recruiting and hiring all junior engineers to be added to the payroll in the next twelve months.* By writing out the exact responsibility to be delegated, you form a clearer picture of precisely what you want to do. Also you will have a record of what you decided and when. Such a record is important if you must delegate a large number of responsibilities.

3. *Choose the person to whom you will delegate.* Half your success in delegating depends on choosing the right person for the task —a person who will do the job quickly and efficiently. Take time to consider who would best do the job. List the available delegates and compare their capabilities for performing the task. Choose the person who seems best qualified for the particular job you have in mind. Avoid choosing an all-around man when you have a narrow, specialized task to delegate. Giving this type job to the all-around many may deprive you of his services at a later date when you need someone for a broader, more comprehensive assignment.

Take pleasure in choosing the person to whom you will delegate a task. For in choosing a suitable person, you give that person an opportunity to prove himself. With every opportunity there is an implied obligation. It is this obligation that leads men to extend themselves to the point where they achieve their objectives. In successfully performing the task the delegate acquires experience and confidence. You and your organization derive the benefit of a job well done. You also gain confidence in your own ability to choose people to whom you can successfully delegate.

4. *Meet the person you've chosen; assign him his duties.* Tell the delegate exactly what you expect of him. Put complex assignments in written form, but be certain to discuss the job fully with the man. Give him full opportunity to question any part of the assignment. Meeting with the delegate is extremely important because you can (*a*) tell the man you've personally chosen him for the assignment,

(*b*) outline the task and the results you expect from him, (*c*) explain the amount and extent of authority you are giving him, and (*d*) answer any questions that arise.

Most people are flattered when their supervisor personally chooses them for a specific task—even though it may be a difficult one. When told of the choice most people will become enthusiastic, giving themselves a better chance of achieving their objective. At this meeting you also have a chance to allay any fears the delegate may have of accomplishing the assigned task. A few reassuring words from you can measurably build the delegate's confidence to a level where he can readily accomplish the task.

5. *Clearly define the authority you are delegating.* For as Prof. Ernest Dale, of Cornell University, wrote, "Probably the greatest single difficulty of subordinates everywhere is the power issue. The boss likes the power of making the decisions himself and he does not like to give it up. For power is something that grows on a person: the more you have, the more you want. As an English wit once put it, jokingly: 'Power is wonderful and absolute power is absolutely wonderful.' "[4]

Tell the delegate exactly what authority you are giving him. Don't try to retain authority by being indefinite about what you do and do not expect the delegant to supervise. This only leads to confusion and ineffectual results. Remember that when you delegate authority you release the energies of your subordinate, giving him a greater chance to accomplish more for your organization. Louis A. Allen, well-known management consultant, expressed it this way: "New research proves that you increase output when you give people *control* of their jobs. Properly applied, this method serves as the strongest form of motivation. Why? Because every person closely identifies *his own efforts* with results secured."[5] Allen also observes that the manager ". . . must learn to do those things that help make people feel they really control their jobs. This literally amounts to putting each man in 'business' for himself. Just talking about it won't do the trick. Clearcut, purposeful management action is needed."

So define the authority you are delegating, and give this authority to the person you've chosen. For if you've followed the first four steps given in this section, the chances are good that the delegate will perform his task willingly and efficiently.

6. *Assign full responsibility for the delegated tasks.* With authority there is responsibility—responsibility for the duties assigned. Be certain to point this out to the delegate. Nothing in life is free—so the man being given authority must assume the responsibility for what he does. As noted earlier, everyone loves power. But the price of power (authority) is responsibility. This is a fact the delegate must recognize and live with.

Responsibility makes men out of boys; it can mature a man in days, making him a dependable, efficient subordinate. So take time to detail exactly what you expect of the delegate. Thus, you might say: "Joe, I'll hold you responsible for the output of Plant A. If the output rises, fine. If it decreases, I'll be calling you in for an explanation. Should I have to call you in, which I hope I won't, I'll expect the facts—not excuses."

The successful administrator knows how to delegate. A recent survey of outstanding middle managers showed that 70 to 80 per cent rated high in their ability to delegate. Since assigning responsibility is an important element in successful delegating, you should give this task your full attention.

7. *Assist the delegate with his first few tasks.* Don't throw him into the task—to sink or swim. Such an approach may lose you a valuable man. Instead, guide him. Show him:

- What results you seek
- Typical methods you use
- Problems he may meet
- Shortcuts that save time

In showing him these procedures remember what Galileo said, "You cannot teach a man anything; you can only help him to find it within himself." Guide a man in his first few tasks and, if you've chosen him carefully, he'll find himself and will serve you well.

What specifics can you help a delegate with? There are many. These include budgets, salary data, planning procedures, profit ratios, tax considerations, etc. A wise supervisor will provide his subordinates with the working data they need. For when the subordinate succeeds, so does the supervisor. And as Dumas wrote, "Nothing succeeds like success."

8. *Take him in on decisions involving his authority.* Let's say you've delegated the operation of a research laboratory to one of

your subordinates. Management decides that it might be wise to expand the lab's function to include product development. Do you simply inform the lab director that from now on he may be responsible for *both* research and development? *No!* The operation of the laboratory is your subordinate's responsibility. To arbitrarily assign new duties to the lab without consulting its director usurps his authority.

Call the director in and arrange to have him discuss with management the advisability of including product development as part of the lab's mission. Even though management may have made the decision, they can gain much from a discussion with the director. He can point out many problems that may have to be solved—manpower, budgets, development facilities, etc. Then, even though management announces its decision as an accomplished fact, the director will at least feel that he has had a chance to take part in the analysis of the move. This knowledge will encourage him to make the maximum effort to build an efficient research and development laboratory.

9. *Tell the delegate how he's doing.* Don't allow him to waste time wondering whether you're pleased with the results he's achieving. *Tell him*—the sweetest of all sounds is praise.

When looking for aspects of a man's performance to praise, don't expect that the job will be done exactly the way you would do it. We're all different. Therefore our approaches to a task will vary. So long as the results are satisfactory, praise the man—tell him he's doing an acceptable job.

You can combine praise with gentle correction and achieve excellent results. Thus, if your subordinate did an acceptable job on one portion of his task, praise him for this. If he can improve another portion, tell him so, *after* praising his good work. The praise will encourage him to work harder to improve his performance in the area in which he is deficient. Should you ever question the worth or advisability of praising a subordinate, think of the last time your boss praised you. Remember the enthusiasm it generated within you? Your praise will have the same effect on your subordinate.

10. *Keep in touch with the delegate's progress.* Don't delegate authority, responsibility, and duties and then forget the man. Such an attitude can only lead to trouble and possible failure. Make

regular checks of the delegate's progress, both with him personally and with others with whom he works. Such a checkup enables you to praise a job well-done and correct minor deficiencies.

Use easily understood yardsticks to measure performance. Thus you might choose units produced, dollars worth of manufactured goods, number of patentable inventions, etc. With yardsticks like these there is little room for arguments or misunderstandings. Making a regular check on subordinates keeps you up-to-date on their progress and performance. It puts you in command of their work—as you should be.

Effective Delegation Can Be the Key to Achievement

Dr. Wernher von Braun, outstanding space-flight engineer, uses delegation to achieve his many successes. Using the team concept in the engineering of a rocket or spacecraft, he delegates many of the management responsibilities to his deputy directors. This permits him to concentrate his efforts on program planning and project engineering. Delegating duties to others is one reason why Dr. von Braun has time to answer detailed technical and scientific questions as they arise. He can also maintain a uniform, well-balanced program structure, leading a team from the conception of a bold idea to its accomplishment.

Many other engineers and scientists rely on delegation to widen their achievements and increase their income. Begin today to use the techniques outlined in this chapter, and watch your accomplishments grow.

REFERENCES

[1] Robert Yarnall, Jr., "The Vital Spot in Management," *Mechanical Engineering*, November, 1960.

[2] Keith Davis, *Human Relations at Work*, 2d ed., McGraw-Hill Book Company, Inc., New York, 1962.

[3] Donald A. Laird and Eleanor C. Laird, *The Techniques of Delegating*, McGraw-Hill Book Company, Inc., New York, 1957.

[4] Ernest Dale, "The Power Problem in Delegation," *Think*, March, 1961.

[5] Louis A. Allen, "How to Get a Man to Really Work Hard," *Management Methods*, February, 1960.

HELPFUL READING

Famularo, Joseph J. *Supervisors in Action,* McGraw-Hill Book Company, Inc., New York, 1961.

Hegarty, E. J., *How to Build Job Enthusiasm,* McGraw-Hill Book Company, Inc., New York, 1960.

Maynard, H. B. (ed.), *Top Management Handbook,* McGraw-Hill Book Company, Inc., New York, 1960.

Uris, Auren, *The Efficient Executive,* McGraw-Hill Book Company, Inc., New York, 1957.

20

Speaking and Listening Are Important

> Discretion of speech is more than eloquence; and to speak agreeably to him with whom we deal is more than to speak in good words or in good order.
>
> FRANCIS BACON

Two prime skills that will contribute much to your professional success are speaking and listening. While you might achieve much in pure research without these skills, you cannot communicate effectively unless you've developed some rudimentary speaking and listening skills. All of us in engineering and science have sat through boring presentations of important papers in our field. Every professional society recognizes the enormous waste of time such speeches cause. That's why most societies exhort speakers *not* to read their papers, but to summarize them in an informative, conversational way that will get ideas across to listeners.

Speaking skills are valuable, of course, in many other activities. In conferences, meetings, sales presentations, interviews, and planning sessions the ability to speak well can mean the difference between acceptance and rejection of your ideas. As you rise in your profession, you'll find that more and more of your time is spent in communications. Thus, C. L. Brisley, Ph.D., of Wayne State University, found in a survey [1] that top plant managers spent 80 per cent, or 108 hours per month, talking. Operating managers spent even more time talking—84.6 per cent or 114.4 hours per month. With this much of your time spent in oral communication, it is easy to see why speaking ability can be so important.

Listening skill is as important as speaking. For as Wilson Mizner, well-known American humorist observed, "A good listener is not only popular everywhere, but after a while he knows something." You truly learn when you begin to listen attentively. Not only does listening give you a chance to hear another man out and learn something, you also have an opportunity to observe his speaking habits. Listening gives you an opportunity to observe, judge, and evaluate. While you're talking you have far less chance to perform any appraisals.

Speak to Convey Information

Engineers and scientists attending a meeting of any kind come to acquire information. Certainly, they enjoy a relevant humorous story that sets a scene or illustrates a point. But the fewer the jokes the better. What the average engineer seeks are concrete, usable facts. He wants to share in your experience so he can broaden his own knowledge of your subject. In almost every case the engineer or scientist comes to listen and learn. Therefore you begin with his sympathy and willingness to cooperate.

To get the most across to other engineers and scientists: (1) Concentrate on your technical facts. (2) Forget flowery descriptions, big words, empty platitudes. (3) Give plenty of useful facts—like operating costs, production schedules, manpower requirements, efficiencies, etc. This is the kind of information that keeps the note takers busy. (4) Present your facts in a step-by-step fashion. Begin with the necessary fundamentals and proceed to the pertinent details. (5) Strive for accuracy and precision. Don't try to impress your audience with the depth of your knowledge or the extent of your vocabulary. Seek, instead, to convey as much useful information as possible. You make a far better impression when you convey useful information than when you try to startle people with the intricacy of your subject. Speakers who give useful information are sought out again and again by other engineers and scientists seeking assistance. (6) Never be ashamed to let your enthusiasm show. The same goes for your sense of conviction about your subject. Enthusiasm and conviction build confidence in your listeners. Your ideas will be much more willingly accepted.

When you talk to businessmen, technicians, and others whose knowledge of your subject is very limited, you must use other techniques. For people of these backgrounds: (1) Try to express your ideas in the listener's language. He comprehends this best. (2) Steer clear of your professional jargon whenever possible, unless your listeners are familiar with it. (3) Express your ideas clearly and accurately—remember that it is your ideas that the listener carries away with him. He'll value your ideas far more highly than any momentary impression you may make on him with humorous remarks or astounding gestures. (4) Try to emphasize benefits, savings, efficiencies, and other advantages to your listener. Remember that he's there to learn how you or your ideas can help him.

Professional-paper Presentation Requires Care

As an engineer or scientist headed for greater success and achievement, you'll probably present a number of professional papers during your career. You can score high by preparing properly. Members of learned societies are subjected to so many poor presentations of papers that outstanding deliveries are remembered for years. Some societies even award prizes or certificates for superior presentations. These are usually termed "speaker's awards." They are given in recognition of the fact that no matter how well a paper is written, much of its punch is lost if the author makes a poor presentation to an audience.

Suppose you work hard to learn how to present a professional paper in an effective and interesting way. What good will it do you? The answer is simple: As a speaker you'll get more deserved recognition for work you've done and knowledge you've gained. And if other authors acquire speaking know-how, you'll find life in an audience a lot easier to take.

In addition, there's more to this than just presenting talks to professional societies. "Writing and talking are the most common engineering operations," said Phil Swain, while editor of *Power* magazine. "Other things being equal, skill with words will add between $20,000 and $100,000 to an engineer's lifetime earnings." This is also true for scientists. So the work you put into better presentation of professional papers will pay off now and later.

Eleven Pointers for Better Paper Presentation

You can increase your speaking skills by keeping the following eleven important ideas in mind when presenting any professional paper. Though some are simple, they are often overlooked.

1. *Know why you're talking.* You present a professional paper because you've done something new or interesting, or because you know more about a subject than your fellow engineers or scientists do. You want to convey useful information, for the benefit of your profession and to increase your own professional standing and prestige. Such an exchange of information is basic to all engineering and scientific progress.

The key thought here is the *conveying of useful information. This is why you're talking.* No matter how valuable the work done, your whole presentation falls flat unless you, as the speaker, succeed in transferring your ideas and knowledge to the people in the audience.

If you fail to convey useful information, not only have you wasted your time—more important, you've wasted the time of everyone who came to hear you. As a speaker you have a very real obligation to your listeners. They expect to be instructed, and they have a legitimate complaint if the speaker fails them. Too many engineers and scientists act as if the audience had no choice but to listen. So know why you're talking—*to convey useful information.*

2. *Keep your story simple.* Your audience is usually composed of people having a broad range of interests and capabilities. If you direct your story at the advanced segment of your audience, you will lose the interest of the less knowledgeable. Since most of your audiences will be composed of more people who've come to learn than of specialists, you'll convey more information if you keep your story simple.

Keeping your story simple doesn't mean that you must strip your paper of its true professional worth. Instead, it means that you should present your facts in an easy-to-follow sequence. Thus, you might begin with (*a*) what was sought, and (*b*) means available to obtain this information. From here you might proceed to (*c*) tests conducted, (*d*) findings of each test, (*e*) results of each test, (*f*) significance of the tests.

A simple story is one that proceeds from the known to the unknown. See that you follow this line of thought every time you present a professional paper.

3. *Use good illustrations.* Don't plan on using every illustration in the published paper as a slide for your presentation. Instead, choose a few key illustrations that demonstrate the main conclusions of the paper. While there aren't any fixed rules, you should seldom use more than twelve slides in a presentation for which the allotted time is twenty to thirty minutes. Probably six or fewer slides is a better number for the usual technical or scientific paper.

Color slides are becoming more popular every year. The color helps you tell a better story and keeps your audience alert. If you decide to use color slides, try to alternate them with a few black-and-white slides. This contrasts the two types of illustrations and keeps the audience more interested.

4. *Choose the best way of presenting your paper.* Let's say you've done a good job of writing the paper (Chapter 9). Next you must decide how to present it.

There are two immediate choices. First, you can *read* the paper. Second, you can give it from memory, using suitable notes to guide your thoughts. Reading a paper is almost certain to kill interest and put your audience to sleep. You want your talk to convey information to your listeners. *Don't read—it's certain poison.* Use, instead, a "speaking version" of your paper, described in item (5) below.

5. *Prepare a "speaking version" of your paper. Write* the speaking version. "But," you say "I've already written the paper." True, but the speaking version differs in several major respects. First, it should be much shorter than the written version. Your paper includes all details for a complete exposition of your subject. But, in speaking your time will be limited by the schedule of the program.

Second, your time is limited by what the audience can take. Making your presentation too long or filling it too full of details defeats your purpose because you lose your audience.

So limit the speaking version to the key points of the paper—the six topics given in item (2). Don't write the speaking version in sentence form—use clauses and phrases instead. Then there's less chance of you *reading* the speaking version.

6. *Always remember your audience.* Be simple and direct in what you say. Try to express your ideas so clearly that your audience will be impressed with your ability to communicate clearly. *Don't try to impress the audience with the profundity of your knowledge or the extent of your vocabulary.* Remember that every listener has a handicap compared with a reader of your paper. If a listener misses a point, he can't go back and clear it up the way a reader can.

So clarity of expression rates top importance. Keep a sharp eye on your choice of words and the pace of your presentation. This doesn't mean the speed with which you speak (although that is important too), but the rate at which your ideas follow one another. Space your ideas with transitional phrases and clauses. Then your audience will be able to follow ideas as your presentation progresses. Choose transition words and phrases that make the structure of your ideas clear and warn the audience of a coming change in subject.

Keep in mind that speaking is different from writing. Spoken English is loose, informal, colloquial. It uses slang and is lively, colorful. So in the speaking version of your paper express your thoughts in the same language you'll use to the audience. Leave some room for unrehearsed comments and for extra comments about the illustrations you present on slides.

7. *Have your notes ready.* Type the speaking version on 3 x 5-in. cards. Then you'll appear to be speaking from notes. Next, carry the cards with you for several days. Read them at odd moments; get to know the words and thoughts you'll present; drill into your mind the order of the thoughts. But do not try to memorize the data on the cards.

If you follow the above recommendations conscientiously, the cards will work like an extensive set of notes. You'll be able to speak with only occasional glances at the cards. And having the whole story at hand does wonders for your confidence. Since the notes were prepared to sound like your conversation, an aside here and there during the presentation can create the illusion of an extemporaneous talk.

Some experienced engineers and scientists who've presented a number of papers prefer to underline important sections of their paper in red, instead of using file cards. The underlined sentences serve as a topic outline. While this method is fine for *experienced* speakers, it is dangerous for novices because many beginners get stage fright. Then, instead of using the underlined sentences as an outline they read the entire text. Their voice deteriorates to a hushed monotone and the audience dozes. So play safe—use file cards for your first few papers. Once you have enough experience you can switch to the underlining method if you wish.

8. *Don't speak at—talk with.* Strive to get your ideas across—to convey useful information. That's what makes good conversation interesting. Lack of it makes many a talk extremely dull.

Put yourself into the presentation. The ideas you're giving are yours—they're interesting and exciting to you. Don't act as if they're remote and impersonal. Let conviction and enthusiasm show as clearly as they would in a face-to-face conversation. Such natural animation and sincerity help create better communication with your audience.

One big caution: *Take your time.* You don't have to speak slowly but it pays to pause between thoughts. This gives you a chance to refer to your notes to see if you're still on the track. Too many speakers are afraid to pause. They either rush along or fill the gaps with unintelligible grunts.

Speak clearly and loudly enough so all can hear. If there is a mike, speak directly into it. Try to stay the same distance from the microphone throughout the talk. If you turn your head to look at a slide, move the microphone so your voice doesn't fade. Never make any remarks that cannot be heard by everyone in the audience. If you do, the audience will become confused and inattentive. Look at the audience. That way you can tell if they fall asleep.

9. *Make sure you have the needed equipment.* The chairman or vice-chairman of the meeting, if he's on his toes, will see that you have all the facilities you need to make a good presentation. But you must tell him what you need. And, since he may be new to the task, it's a good idea to check the facilities yourself.

Is the rostrum in the right place? Can you refer to slides without standing in front of the screen? Is there a pointer handy? Is there a reading lamp so you can see your notes when the lights are out? Does it work? If you need a blackboard, is it in the right spot and equipped with chalk and an eraser?

For a good presentation consider that you and the chairman are putting on a show. To save it from being a flop, you must review every step of the presentation with him. Make sure you're both working from the same scenario and that you know your roles.

10. *Watch out for booby traps with slides.* Good slides add much to the interest and value of your presentation. Poor slides can ruin it.

Technical and scientific societies have pamphlets on preparing and presenting slides, or you can refer to your library for a book giving these data. One book [2] listed in the bibliography of the present chapter gives useful rules for size of lettering and other factors involved in making good slides. Study these rules and use them. Never begin a paper presentation without first looking at your slides as projected on a screen.

Make sure the operator's thumbmark is in the correct spot on the slide border and that the slides are in the correct order. Check with the projector operator *before* the meeting. Find out if the projector is properly set up and focused. Tell him how many slides you have, what order they're in, and about where they come in the talk. Don't leave anything to chance.

If you want to refer back to one or more slides after they have been shown once, tell him about it. Remember, he's working in the dark. He'll have a tough time finding slides 3 and 7 if you spring it on him without warning. Also check the lights and arrange with the operator as to how they should be handled.

Try to make the slides an integral part of your talk. Don't put your listeners to sleep with a dull, slideless presentation and then try to wake them up at the end. You won't transmit useful information by rushing through fifteen slides in quick succession when time is short. It takes time to absorb information.

11. *Be ready for audience discussion.* After you've presented your paper, you are still not quite through. At most professional meetings there will be discussions of your paper. How this is handled depends to some extent on the rules of the society and the wishes of the chairman.

If you can answer each comment as it's raised, there is no particular problem. But if the chairman wants to have all the discussion at once, to be answered in one closure, you have some work to do.

Be sure you have paper to take notes. Jot down the points and questions raised by each discusser. Get the discusser's name so you can refer to him correctly.

When you are called on to answer the discussers, take up each comment in turn and dispose of it to the best of your ability. Your answers must be extemporaneous, whether you like it or not.

Listening Skill Pays Off

Earlier we saw that many managers spend more than 80 per cent of their time talking. Other estimates [3] show that talking accounts for 90 per cent of business communication. While there is much talking in business, there is a lot less listening. For example, Dr. Wesley Wicksell, professor at Louisiana State University, estimates that the majority of supervisors are only 20 per cent effective in listening to their men.[3]

Attentive listening will pay off for you in at least seven ways. (1) You save time because you obtain more information while listening than when talking. (2) With more information you can make quicker and more accurate decisions (see Chapter 18). (3) Your understanding of a given problem or area is increased. (4) You acquire a better knowledge of the person speaking when you devote your full attention to him. (5) Your listening encourages others to tell you more of what they know. (6) By listening after you speak, you can quickly detect if you've been understood. (7) You acquire more useful ideas that can help improve your creativity (Chapter 5) and your problem-solving ability (Chapter 6).

Improve Your Listening Skills

You can boost your listening comprehension by 25 per cent, or more, by training yourself. Listening is improved by thinking in terms of the speaker's objective, weighing his evidence, searching for other clues to meaning, and reviewing the facts presented. Listening is a conscious, positive act requiring willpower. It is not a simple, passive exposure to sound. To listen attentively you must concentrate forcefully on the message in order to keep from daydreaming or letting your mind wander to another subject.[4] Here are seven hints for improving your listening skills during small meetings.

1. *Don't talk—listen.* You can't listen while you talk. Two people talking at once waste each other's time and don't communicate. So close your mouth tightly, and keep it closed while you're listening. This takes will power but you can learn by constant practice. Begin, during your next interview, to practice closing your mouth and listening.

2. *Relax and have the speaker do the same.* Make the speaker anxious to talk by telling him that you're ready and willing to listen. Put yourself in his place for a moment. You'll quickly realize that a cordial attitude on your part will build his confidence and encourage him to say more.

3. *Give the speaker your full attention.* Don't glance at the papers on your desk, look out the window, or watch the clock. Look the speaker full in the eyes. Show him by the expression on your face and by what you say that you want to hear what he has come to say to you.

4. *Eliminate interference.* Is there too much noise in the room? Can others overhear what the speaker says? Is the room too warm or too cold? Is there a light shining directly in the speaker's eyes? All these interferences can reduce a man's desire to talk. Try to choose a quiet, private, cool, and properly lighted room for important discussions. Such an atmosphere will permit you to hear more in less time. It will also encourage the speaker to be more direct and to reveal more information.

5. *Listen and control yourself.* Hear the man out. Don't interrupt. Allow him to express his thoughts, ideas, complaints. If he begins to wander off the topic of the meeting, gently direct his thoughts and talk to the main point. Control your emotions—never blurt out in anger—you will frighten the speaker. Or if something the speaker says seems humorous, refrain from smiling, unless he's trying to be funny.

6. *Draw the speaker out with questions.* Key your questions to his statements. Try to use some of his words in your questions. This indicates how attentive you are and spurs him on to tell you more.

7. *Don't disagree until after the speaker is finished.* If it is necessary for you to disagree with the speaker, do so in a friendly and

considerate way. Try to find some favorable aspects to his thoughts or suggestions. Then gently show him why you disagree and what he can do to change his thinking. Never bang on your desk or make other loud demonstrations of disagreement. Keep in mind at all times that the speaker may be frightened of you, your position, your office, or any number of other aspects of the situation. You can be firm while quiet, a leader while listening, and friendly while helpful. Be ready to listen again, after presenting the reasons for your disagreeing. As Alfred M. Cooper, well-known business writer says, "Nature gave man two ears but only one tongue, which is a gentle hint that he should listen more than he talks." [5]

Control Your Listening Skills

Building good listening skills is important. But you can overdo listening. Once people know you have a sympathetic ear, they may try to monopolize your time. When this occurs, you must disengage yourself by concluding the interview.

When the speaker gets off the track or turns to personal items, it's time for you to give some final directions, opinions, or advice. Then gently indicate that you have other work to do. Knowing what and how much to listen to is part of your listening skill.

Never listen to gossips and scandal-mongers. They will waste your time and poison your thoughts. You can be a good listener without getting yourself involved in useless, debasing chatter.

Professional Meetings Require Listening Skills Too

You can profit immeasurably if you learn to listen more effectively at professional meetings.[6] The average meeting at which papers are presented is a real test of your listening skills because there are so many distractions—people entering and leaving the room, excessive noise, high or low room temperature, whispering between members of the audience, a boring speaker, and your own daydreams. Since a professional meeting gives you an opportunity to hear the results of important work done by others in your field, you should make every effort to listen more attentively to what is said. Here are five useful hints for improving your professional-meeting listening habits.

1. *Read the paper before the meeting.* Then you'll know what to listen for. Coming to a meeting without having read the paper puts you at a definite disadvantage. When you must attend a meeting without having first read the paper, try to quickly scan the paper during the first few minutes of the meeting. This will give you a good idea of the high points of the paper and the facts you should listen for. Since the first few minutes of a paper presentation are seldom devoted to key facts you probably will not miss any important statements. Three minutes spent scanning a paper can triple your comprehension of the speaker's presentation, if you listen attentively.

2. *Keep alert for the speaker's main points.* Don't allow your mind to wander—you may miss the speaker's main conclusions. To keep your mind from wandering during a boring talk, try to relate what the speaker says to your own professional practice. This will give you a respite from the boring talk but will keep your mind occupied with your professional work and enable you to relate what you hear to your activities.

3. *Make notes of important points and questions you wish to ask.* Notes will help you recall what you hear; the act of writing down what is said helps prevent your attention from lagging. Often you'll find that notes can be made directly on your copy of the paper. Then there's less chance of losing either. By writing your questions on the paper you can relate each to a specific portion of the paper.

4. *Participate in the discussion.* Ask questions—and *listen* to the answers. You'll learn more by taking part in the discussion than if you just sit by and idly listen to others discuss the paper. What's more, you'll have to listen to the replies of the speaker, if you wish to know the answers to your questions. After the discussion is over, write a brief summary of what you heard so you'll remember it longer.

5. *Practice good meeting tactics at all times.* Thus (*a*) stay out of drafty corners; (*b*) get as far away as possible from attendees who are trying to hold a meeting within a meeting; (*c*) be prepared—have a copy of the paper with you; (*d*) choose a seat giving a clear view of the speaker, blackboard, and projection screen, and (*e*) have a pencil or pen ready, obtain an ashtray if you smoke, concentrate on what the speaker is saying.

Develop your listening skill, and you will be better prepared to achieve more in all areas of your professional and personal life. Keep in mind at all times the wise observation of Clint Murchison: "You ain't learnin' nothing when you're talkin'."

REFERENCES

[1] C. L. Brisley, "8 Tips to Help You Save Time," *Factory Management and Maintenance,* December, 1958.

[2] Tyler G. Hicks, *Writing for Engineering and Science,* McGraw-Hill Book Company, Inc., New York, 1961.

[3] Dennis Murphy, *Better Business Communication,* McGraw-Hill Book Company, Inc., New York, 1957.

[4] Keith Davis, *Human Relations at Work,* 2d ed., McGraw-Hill Book Company, Inc., New York, 1962.

[5] Alfred M. Cooper, "The Art of Good Listening," *Manage,* February, 1959.

[6] Ralph G. Nichols and Leonard A. Stevens, *Are You Listening?* McGraw-Hill Book Company, Inc., New York, 1957.

HELPFUL READING

Dietrich, John E., and Keith Brooks, *Practical Speaking for the Technical Man,* Prentice-Hall, Inc., Englewood Cliffs, N.J., 1958.

Kruger, A. N., *Modern Debate: Its Logic and Strategy,* McGraw-Hill Book Company, Inc., New York, 1960.

Loney, G. M., *Briefing and Conference Techniques,* McGraw-Hill Book Company, Inc., New York, 1959.

Murphy, Dennis, *Better Business Communication,* McGraw-Hill Book Company, Inc., New York, 1957.

Price, S. S., *How to Speak with Power,* McGraw-Hill Book Company, Inc., New York, 1959.

Surles, Lynn, and W. A. Stanbury, Jr., *The Art of Persuasive Talking,* McGraw-Hill Book Company, Inc., New York, 1960.

Whiting, P. H., *How to Speak and Write with Humor,* McGraw-Hill Book Company, Inc., New York, 1959.

Price, S. S., How to Speak with Power, McGraw-Hill Book Company, Inc., New York, 1958.

21

Build Your Interviewing Abilities

> Because people are people, the seemingly simple process of face-to-face contact isn't simple. . . .
>
> AUREN URIS

The higher you rise in engineering or science, the more time you must spend talking with people. These people may be reporting facts to you, seeking employment, consulting on a problem, or seeing you for hundreds of other purposes. Regardless of the content of the meeting you and the other person are meeting to accomplish a specific objective. Since any meeting requires an expenditure of time, the more rapidly you achieve your objective, the more time you will have available for other work. For this reason developing your interviewing techniques will save you time and enable you to achieve more through and with other people.

Plan Important Interviews

Many engineers and scientists try to conduct important interviews by intuition—without any advance planning. This technique seldom works because you waste time floundering for results. Every important interview rates some planning. Difficult interviews require more planning than simple ones. Once you acquire the habit of planning interviews, you will be able to rapidly organize your thoughts for any upcoming meeting. While this planning will require a few minutes, good plans may save an hour or more during a meeting.

Write out the purpose of your interview. Make the purpose as concise as possible. Typical purposes for which you might hold interviews include (1) employment application—in which you determine the suitability of an applicant for a given position; (2) data acquisition—in which you collect facts pertaining to a given problem within or outside of your organization; (3) data evaluation—in which you and one or more of your associates weigh and evaluate facts important to your organization; (4) decision making—in which you and your associates choose a course of action to achieve a specific objective; (5) problem solving—in which you and your associates develop ideas to solve an important problem.

There are, of course, many other purposes for which you might conduct an interview. When writing out the purpose of any interview, try to use a general description followed by a specific statement of the goal of the interview. Thus, for an employment interview you might write the objective as: *Evaluate employment applicants to fill the post of director of research in the new ABC research laboratory.* Note that this statement gives (*a*) purpose of interview, (*b*) title of position to be filled, and (*c*) place where the applicant will be employed. All these facts are important in your evaluation of any applicant. Knowing in advance what you're trying to accomplish in any interview enables you to direct your efforts toward the specific goal you've chosen.

Retain Command of Every Interview

Many people you interview will be tense and nervous because they fear that the results will not be what they desire. Recognize this fact and you will be much better equipped to make people you interview feel at ease. Remember: The successful outcome of an interview is as much in your hands as it is in the interviewee's. So be on the lookout for anger, fright, illness, and other factors that may alter a person's actions during an interview. Keep in mind that you are an adult—and act like one—even though the interviewee may try to annoy or upset you. A calm attitude contributes much to controlling the course of any interview.

When first learning correct interview techniques some engineers and scientists lose courage and give up. They find that tailoring their conduct to that of another person is irksome and uninteresting. If you experience such a reaction, do not become discouraged. For if you continue your study of interview techniques, you will soon find that working with people has infinite variety and offers a constant challenge. If you desire to achieve more in your profession, you will almost certainly have to become adept at interviewing people at all levels in your organization and with many different educational and experience backgrounds.

Be considerate of the interviewee—remember that he is a human being. He has ideas, hopes, skills, opinions. While you should retain command of an interview, you learn little when you do most of the talking. So give the interviewee a chance to talk. Watch him, and listen attentively. A few minutes of careful observation and listening will often tell you more than an hour of questioning. Should the interview begin to wander, bring it back to the subject by gently saying, "Well, we wanted to get this ABC issue settled," or some such remark that serves as a transition from another topic to the subject of the interview. You are in command of the interview—be certain you do not lose control of it.

Experienced interviewers often use a desk clock that is not visible to the interviewee. By keeping track of the elapsed time the interviewer is able to get as much useful information as possible. Remember, however, that in almost every interview you must allow some time for aimless chitchat. Such talk will often help the interviewee relax, and you can then obtain more information from him.

Be on Guard for Interview Pitfalls

During any interview you can err in either of two ways: (1) prejudice against some characteristic of the interviewee, or (2) emotional involvement with the interviewee because of some shared enthusiasm. You must be on guard against both these pitfalls because they can warp your judgment and lead you to erroneous conclusions about an interviewee or the facts he presents.

Prejudice can lead you astray more than you might ever realize. Thus, if you're interviewing research scientists and find that two stout candidates fail to impress you, review your prejudices. Perhaps you are slender and have developed a dislike for stout scientists. While a stout scientist may not have an attractive appearance, this is unrelated to his research abilities. Some of the best researchers —Einstein, Steinmetz, Tesla—would never have won a contest for the most handsome man. Yet each made major contributions in his field.

So check your prejudices—and rid yourself of them. Prejudices can range from physical appearance to school attended, type of clothing worn, previous employers, mannerisms, speech accent, etc. In ridding yourself of prejudices, you become more objective and can judge people and their ideas more accurately. Also you make yourself a wiser human being, better equipped to enjoy knowing and working with other people. For as William Hazlitt said—"Prejudice is the child of ignorance." Discarding your prejudices helps lift the veil of ignorance that clouds the eyes of many of us.

Emotional involvment with an interviewee can start from a chance remark like, "I'm a New York Yankee fan," or "I play golf every chance I get." If you're a fan of this club, or a golf addict, you may suddenly find yourself agreeing with the interviewee before you've given his words much thought. Or if you dislike that particular baseball team or game, you may find that you are strongly disagreeing with the interviewee. Either reaction is an emotional one and should be guarded against.

When you react emotionally to something an interviewee says, discuss the subject with him for a few minutes. By doing so you will learn more about the interviewee and his reasons for a particular penchant or dislike. This information will enable you to judge him and his ideas more objectively. Also, you'll be more ready to return to the topic of the interview. Remember to judge the man, or his ideas, in view of the purpose for which you're holding the interview. Using any other basis will only lead you to questionable conclusions.

Be Fair and Alert at All Times

Don't try to hold back information that might be helpful to the interviewee. When interviewing a man for a position in your organization, try to give complete data about the job. Only when an interviewee has this information can he fairly decide if he can meet the requirements of a particular job. If you hold back information, he may accept the position and decide, after working a few days, that he dislikes the tasks assigned him. Should he then quit, both you and he have wasted valuable time. Or if he asks for a transfer, you have a time-consuming problem on your hands. So be fair with every interviewee. You expect most people to level with you. Treat others in the same way.

Keep alert during every interview. You'll never learn much if you allow yourself to become absentminded or a daydreamer during dull parts of an interview. Words spoken during a lull in an interview may be keys to what direction the discussion will take. If you aren't alert, these key items may pass unnoticed, and you will have difficulty following the line of thought. So form the habit of listening and watching intently during every interview. You'll learn more and will make a better impression on the interviewee. This habit will also be a big asset when someone is interviewing you.

Perfection Is Seldom Found

In any interview you may find yourself seeking perfection—the exact answer, ideal job candidate, etc. While perfection is a commendable goal, it is seldom achieved in any part of life. Yet many engineers and scientists go on seeking the perfect outcome from every interview. The result is that they seldom find what they consider is a suitable job applicant or answer to a problem. This characteristic of some engineers and scientists is one of the causes of friction between these men and businessmen. The latter, recognizing that perfection is seldom attained, are satisfied to work with reasonably well-qualified job applicants or with partial answers to problems. Such operating methods are necessary if the competition offered by other organizations is to be overcome.

Set yourself high standards—but be satisfied with something less than perfection. Don't damn a man because you find one or two flaws in his background, education, or reasoning. No one—not even yourself—is perfect. The sooner you recognize this fact, the easier your life will become. Look for the strong characteristics in a man, for the usable portion of his ideas. Recognize how you can use these, and you and your organization will gain much. As Michelangelo said, "Trifles make perfection, and perfection is no trifle."

Eleven Tips for Better Interviews

You can get more from every interview if you keep these short tips in mind:

- When interviewing a woman, have another woman sit in on a portion of the discussion. Women tend to view one another more objectively than a man does.
- Provide privacy for confidential discussions—close the door.
- Don't argue—listen. If you must disagree, try to agree first—then disagree.
- Have needed supplies handy—scratch pads, pencils, rulers, charts, etc.
- Keep interruptions from outside at a minimum—don't answer any phone calls except long-distance.
- See that everyone has a seat—people tend to be more co-operative when seated.
- Try to root out your prejudices—just because a man has a different outlook from you doesn't mean he's wrong.
- Be ready to sell your ideas to others if they show resistance. An interview often resembles a sales meeting in that you must convince a man to come to work for your organization, accept a new assignment, etc.
- Take time to prepare for important interviews. You'll profit more by doing so.
- Be understanding—some people panic during an interview. By putting them at ease you will obtain useful information or ideas that will make the time spent on the interview worthwhile.
- Don't waste time in interviews with complainers who want to cry on your shoulder, gossips anxious to curry favors, or slanderers anxious to damage another person's reputation. You gain nothing from such discussions and stand to lose your self-respect.

Check Your Interview Techniques

Here's a quick checklist to help you determine how effectively you now conduct interviews. You should score yes on better than half of the questions.

	YES	NO
1. Do you listen attentively while the interviewee is speaking?	____	____
2. Do you try to draw the interviewee out with questions?	____	____
3. Are you understanding of another person's feelings during an interview?	____	____
4. Do you plan the purpose and procedure of important interviews?	____	____
5. Do you try to forget your prejudices during an interview so you can judge results objectively?	____	____
6. Do you retain command of an interview, directing it to yield the information you need?	____	____
7. Are you often swayed by an emotional attachment to an interviewee or to his interests?	____	____
8. Do you seek perfection in people you interview and those who work for you?	____	____

HELPFUL READING

Famularo, Joseph J., *Supervisors in Action,* McGraw-Hill Book Company, Inc., New York, 1961.

Fear, Richard A., *The Evaluation Interview,* McGraw-Hill Book Company, Inc., New York, 1958.

Kahn, Robert L., and Charles F. Connell, *The Dynamics of Interviewing; Theory, Techniques, and Cases,* John Wiley & Sons, Inc., New York, 1957.

Maier, N. R. F., *The Appraisal Interview,* John Wiley & Sons, Inc., New York, 1958.

Uris, Auren, *The Efficient Executive,* McGraw-Hill Book Company, Inc., New York, 1957.

22

Learn How to Run Effective Meetings

> "Oh no! Not another committee meeting," the executive groaned. "It's only Wednesday morning, and I've been to seven meetings already this week. When am I going to get my work done?"
>
> KEITH DAVIS [1]

Today's engineers and scientists probably attend more meetings as part of their professional activities than any similar group of men. Meetings are called for hundreds of different reasons—to make group decisions, brainstorm new ideas, plan future projects, etc. With team effort as an important part of our technical and scientific life we're all required to attend more and longer meetings. Management, recognizing the importance of meetings to achievement in all fields, is placing more emphasis than ever on effective results from every meeting.

Meetings Are Important to You

To achieve more in engineering or science, you must know how to run effective meetings. As you rise in your profession, management *expects* you to be able to run productive meetings. For when you are capable of conducting a useful meeting, you can achieve more by directing the efforts of several people toward a useful goal.

You *can* learn how to run an effective meeting. With a knowledge of the essentials of productive meetings you can practice the important techniques during every meeting you attend. Combining knowledge and practice will soon enable you to direct the efforts of your colleagues smoothly and efficiently. This chapter presents the essentials of conducting effective meetings in all branches of science and technology. Make these techniques a part of your regular skills and watch your achievements soar. Remember: *Capable supervisors know how to run effective meetings.*

Sixteen Techniques for Better Meetings

To succeed, every meeting must have (*a*) strong leadership and (*b*) adequate planning. Leadership and planning are inseparable—the efficient leader plans every meeting as far in advance as possible. He knows that leadership alone accomplishes little without a planned agenda designed to achieve results. Hence, you'll find that many of the following techniques are well-suited to your background and training in logical thinking.

1. *Decide, in advance, on the purpose of every meeting.* Know why you're holding the meeting. Write out your purpose, using as few words as possible—preferably less than ten. Thus, you might write: "Choose powerplant for XYZ spacecraft." Or: "Select new products for manufacture in Plant A." Or: "Plan research schedule and goals for the continuous laser pump."

Knowing the purpose of a meeting will help direct every step you take. It will give you a feeling of confidence, making you a stronger, more effective leader.

2. *Prepare an outline of the topics to be covered in the meeting.* List every item of importance—but don't try to cover 90 topics in five minutes—it can't be done. Choose attainable goals for your meeting. Tailor the outline to these goals. Remember—achieving one worthwhile goal is far more important than several hours of rambling, useless discussion. Keep your outline simple—phrases, clauses, or sentences are usually sufficient topic entries. Thus, you might write:

a. Describe powerplant *selection alternatives* for XYZ spacecraft
b. Discuss *advantages* of each powerplant alternative
c. Discuss *disadvantages* of each powerplant alternative
d. *Choose* best powerplant for XYZ spacecraft

Note how this outline ties in with the first purpose given in item (1) above: *Choose* powerplant for XYZ spacecraft.

3. *Choose and notify, in writing, prospective attendees that the meeting will be held.* State in your notice (*a*) the purpose of the meeting, (*b*) the day of the week the meeting will be held, (*c*) the date, (*d*) the place of the meeting, and (*e*) any special items—reports, equipment, products, studies, etc.—the attendees should bring to the meeting. Be exact in every statement concerning the purpose, time, and place of the meeting. Where you wish to limit the duration of the meeting, state the time limit—one hour, thirty minutes, forty-five minutes, etc.—in the notice. Send the original copy of your notice to the first prospective attendee on your list and a carbon copy to each of the other attendees. To inform every attendee of who will be at the meeting, it is usual practice to list on the notice the names of the other people to whom copies were sent.

Send your meeting notice as far in advance of the meeting date as possible. If you send the notice three or four weeks in advance, ask for confirmation that the attendee will be present. On the day before, or day of, the meeting either send a written reminder of the meeting, or phone the attendee. This procedure will ensure better attendance at every meeting you hold.

Note that you needn't inform nor remind people about regularly scheduled meetings which their duties require that they attend. These meetings are seldom forgotten. The meeting that is likely to be overlooked in the rush of daily activities is the nonroutine one called a week or more in advance of its actual date. You owe it to every attendee to notify him in advance and to remind him of the meeting. Using this procedure will give you the reputation of a competent meeting chairman.

4. *Reserve, in writing, a suitable room for the meeting.* Be sure the room is (*a*) large enough; (*b*) has enough seats; (*c*) has necessary facilities like a blackboard, slide projector, public address system, etc., depending on the size and nature of the meeting.

Remember: There is nothing so embarrassing as trying to hold a meeting in a room already occupied by people for another purpose.

So reserve the room in writing—and insist on getting a confirmation of the reservation in writing. Then you'll be able to approach the meeting in a more relaxed mood. Too many engineers and scientists have been shunted from one room to another seeking meeting space to make this situation humorous. The wasted time and effort can discourage clear thinking and effective effort in the meeting when space is finally found.

5. *Collect supplies needed for the meeting.* Check to see that pads of paper, pencils, reports, statistics, drawings, and other items you and the attendees will need for the meeting are available. Bring a few extra of each item so you will not be caught short if unexpected guests attend. Engineers and scientists have a penchant for strolling into meetings uninvited. These "guests" can interfere with the meeting if they must look over someone's shoulder to read reports, drawings, or other documents.

6. *Check the seating arrangement.* See that there are enough seats, plus a few extra. If people have to carry chairs into the room after the meeting starts, they will disrupt thoughts and discussions. Be sure the meeting chairman is in a seat that gives him a full view of the room and the attendees.

7. *Start your meeting on time.* Remember that every minute wasted in a meeting must be multiplied by the number of people present. Thus, if you waste ten minutes when six people are present, sixty minutes are lost forever. So start precisely on time. When a few known laggards will be attending, call them five minutes before the starting time and remind them that they are due at the meeting. Reminders like this get results.

8. *Begin the meeting on a positive basis.* Don't apologize for calling the meeting—too many engineers and scientists waste time fumbling for an excuse. People aren't interested in excuses—they want to get to the point and then adjourn the meeting. Start by telling the attendees what the meeting is for, except where the meeting is a routine one held at regular intervals for a single purpose. State a time limit. Be sure the subject of the meeting and the time limit you set agree with the notice you sent announcing the meeting. If you change either the topic or time schedule, you may cause confusion. One way of combining these items in an introductory statement that also starts the meeting is: "Good morning, gentlemen. We're

here to choose the powerplant for the XYZ spacecraft. Since all our studies are complete we should be able to make the choice in the hour we've assigned to this meeting."

9. *Seek full participation from everyone in the group.* Ask questions of individuals; ask for opinions. Listen carefully. Stimulate crosstalk between qualified people. Don't be afraid to raise your voice, laugh, or bang the table. Every business meeting is a serious gathering—but this doesn't mean that the room must be as quiet as a morgue. Remember: Silent attendees give little help in solving problems. The silent man may be daydreaming, his mind a hundred miles away. Get him talking—his opinion may provide the answer you are seeking.

10. *Keep the discussion on the meeting topic.* Don't let the talk wander to unrelated subjects. As soon as you hear a man beginning to talk about a topic not concerning the meeting, stop him short with, "That's interesting, Jack. We'll come back to it later, if we have time. At the moment, we want to choose the engine for the XYZ spacecraft. What's your recommendation?" Don't be afraid to aggressively lead an attendee back to the subject of the meeting. By doing so you save his time and the time of every one else at the meeting. Also, by concentrating on the meeting subject you help the attendees devote more effort to solving the problem facing them.

11. *Allow time for full discussion of the meeting topic.* Don't cut a man short when he's making a valuable contribution to your meeting. Give him time to say what he thinks. Your meetings will accomplish more if you make a determined effort to obtain an opinion from every attendee. Concentrate on the silent attendees—many are too timid to venture an opinion, unless asked. Stimulate discussions between attendees. Such cross talk will often bring out new ideas and will help you determine the majority opinion.

12. *Summarize, briefly, as each goal is achieved.* Every meeting has three parts—a beginning, a middle, and an end. The middle is often dull because the attendees believe that little has been accomplished. You can overcome this dullness by summarizing what you've accomplished during the first ten-, twenty-, or thirty-minute intervals. Thus, you might say, "Gentlemen, we've now ruled out two of the five engine possibilities for the XYZ spacecraft. Now let's con-

centrate on the remaining three." Such a statement lends an air of accomplishment to the meeting and encourages the participants to tackle the remaining problems with vigor.

13. *Explain complex ideas—don't lose people's attention.* Complex ideas are certain to arise. Watch the faces of the people at your meeting—you'll soon recognize when a man is puzzled. Explain the idea, fact, theory to everyone present. If, as sometimes happens, you can't explain the idea yourself, call on someone present who can. By following this procedure, you will help all attendees get a better understanding of every idea presented. This will spur fuller discussions, leading to greater accomplishment. Explaining ideas is particularly important among engineers and scientists because one misunderstanding can lose you the opinions of one or more attendees.

14. *End your meeting on time.* Don't allow the meeting to drag on; end it when the allotted time has elapsed. *Summarize the findings, decisions, and actions to be taken.* Be sure everyone understands what he is to do—if specific tasks have been assigned as a result of the meeting. Put your summary in positive terms—don't apologize or mumble. Thus, you might summarize with: "The XYZ spacecraft will use the H-15 liquid-oxygen engine. We will begin production of this engine on June 1. One month prior—on May 1—we will meet to decide on the production schedule for this engine. At that meeting Charlie Smith and Joe Jones will present a comprehensive report on the best production schedule."

15. *Write a brief résumé of the meeting.* Circulate one copy of the résumé to each attendee. Where specific action is to be taken, underline the assigned task. Call attention to this by noting on the first page: "Joe—note p. 2 item for your follow-up." Send a copy of the résumé to your supervisor so he knows what you accomplished at the meeting. Where others in the organization are interested in the subject of the meeting, send them a copy too. This will help publicize the efforts of the group and will keep others informed of the accomplishments made at the meeting.

16. *Take immediate action on any decisions made at the meeting.* Don't be the "think-it-through" type of engineer or scientist. Get to work immediately—as soon as you leave the meeting. Delay at this time will only lead to inefficiency and wasted effort at a later date when you try to remember what you're supposed to do. Also, sooner than you realize, another meeting will be scheduled. If you don't

accomplish the objectives of the first meeting, you'll have little hope making a success of the next.

Avoid meeting pitfalls. Many engineers and scientists have wasted time in poorly conducted meetings. You probably remember two or three of these "spectaculars" where you were bored, itchy, and tense. Adjournment was the most welcome word you heard that day. Let's take a quick look at some ills that may befall any meeting. Knowing these pitfalls in advance will help you avoid them.

1. *The meeting isn't planned in advance.* Subject, time, place, duration, attendees, facilities are neglected. So people mill about without adequate directions or goals. Time is wasted; tempers grow hot; little is accomplished.

2. *The subject is too general to produce useful results.* With a topic like *world-wide defense measures* many meeting chairmen would find it difficult to focus the attention of the attendees. When faced with a topic like this, break it into subtopics like *Atlantic seaboard submarine defense measures, Pacific seaboard submarine defense measures,* etc. These topics are more specific and lend themselves to readier discussion. By finding solutions for a series of more limited topics, the meeting attendees can build to a generalized solution for a broader problem.

3. *Attendees are not carefully chosen.* The topic of the meeting is beyond the understanding or responsibility of most of those present. As a result interest lags, discussion is sporadic, and little is accomplished.

4. *Discussion is allowed to get off the meeting topic.* The full time of the meeting is not devoted to the topic. Decisions may be rushed, preventing adequate study of the available alternatives.

5. *Personality clashes are allowed to engender animosity, distrust.* Technical and scientific meetings must be kept on an impersonal basis. Personality clashes should be squelched as soon as they arise —before the disagreement has time to generate resentment and distrust. Ill will between attendees can discourage discussion and kill creativity.

6. *The meeting chairman allows his authority to be assumed by someone else.* This can lead to the meeting being taken over by one or more persons who are incapable of directing the group. Time is wasted; decisions may not be official.

7. *Supplies and facilities, or both, are inadequate.* Reports, draw-

ings, notebooks, etc., may be missing. Chairs, table, blackboard, etc., may be too few in number or too small. Lights or ventilation is inadequate. Meetings seldom run smoothly in crowded, hot, inefficient rooms. People become logy, inattentive.

8. *Attendees arrive late, interrupt discussions, require filling in about what has gone on.* Urge people to arrive on time. By doing so, they will save themselves time and the meeting will achieve more.

9. *Chairman fails to summarize the meeting.* Attendees leave without adequate knowledge of what was accomplished. The meeting becomes a blur in their minds.

10. *Attendees are not given adequate instructions concerning the action they should take.* So they do nothing. Decisions made at the meeting lie dormant. Time has been wasted—efforts were expended and there are no visible results.

11. *Chairman fails to prepare a résumé of the meeting.* His supervisor and the attendees have no record of what took place, what action was recommended.

Good Meetings Pay Off

You're on the way to greater achievement in engineering and science. One of the keys to your success is learning how to run more effective meetings. Start practicing now, using the hints in this chapter. Since modern engineering and scientific projects are so often conducted as team efforts, the ability to run good meetings is valuable.

REFERENCE

[1] Davis, Keith, *Human Relations at Work,* 2d ed., McGraw-Hill Book Company, Inc., New York, 1962.

HELPFUL READING

Bonner, Hubert, *Group Dynamics: Principles and Applications,* The Ronald Press Company, New York, 1959.

Cartwright, Dorwin, and Frederick Alvin Zander, *Group Dynamics: Research and Theory,* 2d ed., Harper & Row, Publishers, Incorporated, New York, 1960.

Dyer, Frederick, Ross Evans, and Dale Lovell, *Putting Yourself Over in Business,* Prentice-Hall, Inc., Englewood Cliffs, N.J., 1957.

Hannaford, E. S., *Conference Leadership in Business and Industry,* McGraw-Hill Book Company, Inc., New York, 1945.

Hegarty, Edward J., *How to Run Better Meetings,* McGraw-Hill Book Company, Inc., New York, 1957.

Richardson, F. L., *Talk, Work, and Action,* The Society for Applied Anthropology, Ithaca, N.Y., 1961.

Robert, H. M., *Rules of Order Revised,* Scott, Foresman and Company, Chicago, 1951.

Sayles, L. R., *Behavior of Industrial Work Groups: Prediction and Control,* John Wiley & Sons, Inc., New York, 1958.

Stogdill, R. M., *Individual Behavior and Group Achievement,* Oxford University Press, Fair Lawn, N.J., 1959.

Sturgis, A. F., *Sturgis Standard Code of Parliamentary Procedure,* McGraw-Hill Book Company, Inc., New York, 1950.

Zelko, H. P., *Successful Conference and Discussion Techniques,* McGraw-Hill Book Company, Inc., New York, 1957.

23

Get Things Done—Now

> Procrastination is the thief of time.
>
> EDWARD YOUNG

> Never put off till tomorrow what you can do today.
>
> CHESTERFIELD

We engineers and scientists have many admirable traits—like straight thinking, superior analytical ability, well-organized work habits, etc. But one trait many of us seem to lack is the ability to get things done. By nature we seem to be procrastinators. Why? Partly because the usual training of technical personnel emphasizes delayed thinking. Students are told to "given the problem plenty of thought before reaching a decision." This procedure is fine on the campus where life is unhurried, and you'll get a passing grade if the solution is turned in by early June. But in technology and science management is constantly pushing for results. This means that most engineers and scientists must devote some of their energy to learning how to finish the tasks assigned to them.

You *can* learn to accomplish more—to get things done. Your habit of putting things off, allowing your thoughts to jell, can be changed to a habit of completing each task on time or ahead of schedule. How? Let's see.

Be a Self-starter

Develop the habit of beginning routine and nonroutine tasks without having to be told to do so by your supervisor. Don't allow yourself to kill time between jobs. Instead, turn to something productive. Begin, *and finish,* as many assignments as you can. The acts of beginning and finishing will train you in quick anlysis and completion of projects. Some engineers and scientists, believing they are underpaid, adopt the attitude that they will not do anything of their own volition. Such an attitude is harmful in at least two ways. Since these men are not self-starters their supervisors are unwilling to assign them responsible tasks. Because of this these men are held in positions of lesser responsibility. Their monetary rewards are lower, leading to more discouragement and less initiative.

By adopting a self-starting attitude, you will achieve more. And every time you accomplish something new, you will derive a genuine feeling of satisfaction. Also, the successful completion of even the smallest task will give you greater confidence, helping you to achieve more in the future. So stop having a niggardly attitude toward your profession, your job, your organization, and your professional society. Replace this attitude with a willingness to do more on your own. Once you do, your professional accomplishments can zoom.

Go Straight to the Target

Don't be deflected—once you begin a task, finish it. Know what you want to do and how you will do it. Keep pushing until you accomplish your aim.

Some engineers and scientists shun aggressiveness, feeling that such an attitude is unprofessional. While aggressiveness may not appear to be desirable in the professional man, you will accomplish little in life unless you set up goals and strive for them. So do not be afraid of being somewhat aggressive. You will accomplish more with less effort. As the president (an ex-engineer) of a leading aerospace firm said, "Give me engineers and scientists who are a trifle 'hungry' for recognition and compensation. With such a team I can out-design, out-bid, and out-manufacture anyone in the field.

Not only will the firm profit—every one of those engineers and scientists will benefit too. Each will have a larger paycheck and a greater feeling of accomplishment."

Build Up Steam Early

Develop an enthusiasm for completing your task—no matter how small or large it may be. Visualize the results. Picture the completed task and its benefits to your organization. Imagine how your supervisor will react to a job well-done. Review the advantages you will derive—in terms of recognition, a promotion, greater earnings, etc. Of course, you can't expect these returns from every task you accomplish, particularly smaller ones. Even so, good performance on a series of small tasks can win you a favorable reputation that will lead to bigger tasks. So don't, like some engineers and scientists, loaf along at a tenth of your capacity. Build up steam to achieve more—and keep the steam up throughout your career.

Know Yourself

Self-knowledge is important to every engineer and scientist. For the better you know yourself, the less chance there is that you will fool yourself into wasting time and energy. Equally important, the more you learn about yourself, the deeper will your understanding of other people be. And every engineer and scientist can benefit from a better understanding of other people.

So analyze yourself, your desires and goals in life. Know what you want to accomplish and how. With definite goals in view you will find it easier to start and sustain every activity.

Keep Pushing

Don't give up too early. Persevere—you can accomplish most of what you try if you keep trying long enough. Some engineers and scientists lose their drive before they've made enough of an effort. Think back on your last few major tasks. Try to remember if you felt like giving up just before you finished the job. If you did feel like dropping the task but didn't, you probably have enough drive. But if you didn't finish these tasks, if someone else had to, you probably could use more drive.

Before starting a major task, make a time estimate of how long it will take. Don't give up until you've expended at least as much time as you estimated. And before you give up, estimate your chances of success. Usually, you have a greater opportunity of accomplishing your objective if you continue working than if you give up altogether.

Give Yourself Incentives

Reward yourself whenever you complete a difficult task. What kind of an incentive should you use? This depends on your interests. One mechanical engineer, who is studying electronics in his spare time (and enjoys it), rewards himself for another job well-done by giving two hours of his time to electronics, instead of the one hour he normally spends on study. Thus, he accomplishes two aims—(1) he gets a job done, and (2) he learns more about an important subject.

There are a number of variations to the incentive scheme. A chemist, whose hobby is boating, buys some needed equipment for the boat whenever his efforts deserve a reward. A technical writer takes his wife to a Broadway play; an electronics engineer takes a full day off from work of all kinds and listens to hi-fi music. Choose your own reward, and have it waiting when the job is finished. An incentive will spur you to greater efforts and make your task more interesting.

Generate Enthusiasm

Some engineers and scientists try to hide their enthusiasm for a task. They think that a disinterested air shows sophistication. Actually, enthusiasm is one of the traits most desired by any supervisor—be he president, vice-president, chief engineer, project engineer, or manager. Discard the idea that enthusiasm is childish—it never was and never will be.

Generate your enthusiasm by finding the beneficial aspects of your task. Try to find features of the task that appeal to your technical or scientific interests. Don't bridle your interest and enthusiasm. Allow them to build to a high pitch. For as Emerson said, "Nothing great was ever achieved without enthusiasm." Properly kindled and controlled, enthusiasm can help you get things done and increase your professional achievement.

Be Creative

Develop new ways to accomplish routine tasks. Be creative about what you do, and how you do it. One aeronautical engineer says, "I'm number-happy. I number the steps in every task I have. Then as I finish each step, I cross it off. This technique keeps me interested, gives me a feeling of satisfaction—*and gets things done.*" A marine engineer who read this statement said:

> I'll do him one better. All my life I've tried to be more creative about my duties. During a recent war when we had to quickly train thousands of young inexperienced men to operate the power plants of merchant ships, I prepared an entire book that consisted of numbered steps to be followed in starting, maneuvering, and stopping a ship's engines and boilers. Some engineers objected to this technique. But it worked—hundreds of young engineers told me that the numbered steps made them think through their jobs like they had never done before.

Creative people get more done in less time. So develop your creativity, and achieve more. See Chapter 5 for hints on developing your creative abilities.

Aim at Quality

We all know engineers and scientists who accomplish much. But the quality of their work is so low that they are wasting their time. Getting things done is easy, if you don't aim at quality. When you raise your sights and try to achieve a consistent level of quality, your tasks become more difficult. An enormous low-quality output can be worthless. So decide today to settle only for top-quality work. Insist on this level of work from yourself and your associates. Your achievements and contributions will rise everytime you raise your quality standards.

Be Positive, Persistent, Productive

Look for the positive aspects of each task. Don't, like some engineers and scientists, continually tear down tasks with criticism and negative attitudes. Persistent criticism of your supervisors, your organization, and the work of your group can poison every aspect of your life. You will find that your associates shun you, that your supervisors will assign you only routine tasks because your negative attitude is discouraging and unpleasant. Some engineers and scientists believe that a negative and critical attitude shows how "professional" they are. Actually, just the opposite is true—this attitude is characteristic of the immature person.

So look for the positive aspects in every project. As one civil engineer reports:

> When I started looking for the good in people and tasks, a new world opened to me. I began to see that even the most humble tasks—like delivering blueprints from one office to another—have a purpose and importance to mankind. With this understanding came a sympathy for all people who make a useful contribution to the world. I am now a more useful and creative person because I understand and sympathize with others—instead of wasting my energies trying to criticize them.

Be persistent in your efforts to get things done. You won't accomplish much in your profession unless you strive to complete every task you begin. Successfully finishing every task you start develops a favorable atmosphere in your mind. You acquire the courage to begin and finish new, more difficult assignments. Persistence pays dividends in every phase of engineering and science—begin to develop this desirable trait today.

Be productive—try to produce more than any of your associates. Never limit your output to the "going rate," as some engineers and scientists do. You are a professional man—your skills are valuable to mankind. Limiting your achievements to a level arbitrarily set by someone else is false and harmful. Set your own production pace, no matter what your particular specialty is. Then produce at this pace, making sure that your output is of the highest quality possible.

Encourage Yourself

Keep your enthusiasm at a high level by reading inspirational material like biographies of successful engineers and scientists. One nuclear engineer recommends that "engineers and scientists study proverbs and famous quotations." Usually short, these can stimulate enthusiasm and thought. Studying a few helpful quotations from one author can start you on a reading program that covers all his work. You not only expand your knowledge—you also build a backlog of helpful quotations for those dull, dreary days when you need a lift.

Control Yourself

Your achievements will be minor unless you learn how to control yourself. You must be able to exert enough control to turn your back on pleasures whenever work is necessary. For if you delay, if you take too much time out for leisure, your achievements will shrink. This doesn't mean you must spend all your time working. Instead, it means that you must be able to switch your attention from leisure to work and back again without delay. Having this ability will enable you to start fast and continue your activity until you finish your tasks.

Practice self-control. Assign yourself some difficult tasks—like studying a new subject, working for your professional society, etc. Then force yourself to meticulously keep the commitments you have made. The self-control you learn under these conditions will enable you to achieve more in the important areas of your profession. Self-control is a major asset in the success of every engineer and scientist. Remember what Goethe said: "Self-conquest is the greatest victory."

Do Distasteful Jobs First

Every man in a profession has some irksome tasks. The usual approach to these tasks is to put them off until the last. When you do this, you have a subconscious hope that putting off a task will make it go away. But as your own experience shows, few unpleasant tasks ever go away. They have a habit of hanging on, haunting you until they are done.

Next time you have an unpleasant task, do it first. Use the tasks that follow the unpleasant one as an incentive to begin early, finish fast. By using this technique you may develop a liking for the irksome task. But even if you don't, you will get the bothersome chore out of the way. Once it is done, the thought of it will no longer spoil your day. As Douglas Jerrold wrote: "The ugliest of trades have their moments of pleasure." Get your troublesome duties done first, and then you too can have your moments of pleasure.

Force Yourself

Don't expect to achieve more in today's engineering and science with only a half-hearted effort. Force yourself to do more on your job, in your professional society, in your community, and in your spare time. Practice giving fifteen or thirty minutes more to each important task you undertake. These few extra minutes can mean the difference between a routine output and a major contribution. Samuel Johnson, who knew from personal experience, wrote, "Great works are performed not by strength but by perseverance."

Try forcing yourself for the next month. Keep tabs on your output. If you regularly insist on doing a little more each day, the quality and quantity of your output will improve. You will truly begin to achieve more in your chosen field.

Handle Trifles Quickly

You probably meet all sorts of trifles that tend to bog you down in your work. Your secretary takes a day off, important drawings are misplaced, there are minor errors in a specification, etc. Learn to deal with trifles quickly and efficiently. Do this by analyzing the alternatives open to you. Choose the alternative offering the most effective solution and go on to your next major task. Don't waste time and energy on trifles. Take the best action you can—don't fret or stew. You gain nothing from worry.

Find Your Most Productive Hours

You're probably more productive at certain times of day than at others. This is true of most engineers and scientists. Find your most productive hours by reviewing in your mind the time of day when you feel most creative. You may find that your greatest creativity occurs during the morning, afternoon, or evening hours. Once you've determined your most productive hours, arrange your work schedule around these hours.

One well-known mechanical engineer is most creative between 10 A.M. and noon. His typical daily schedule is:

9:00–10:00 A.M.:	Read mail; plan replies; have coffee
10:00–12:00 noon:	Dictate correspondence, reports, surveys
1:00–3:00 P.M.:	Conferences and meetings with engineering personnel
3:00–4:00 P.M.:	Special phone calls, correspondence, meetings
4:00–5:00 P.M.:	Business and technical reading (magazines, journals, etc.)

By following this schedule as closely as possible, this engineer achieves more every working day. You can arrange a similar schedule for yourself, once you know your most productive hours.

Kid Yourself—and Get More Done

One outstandingly successful engineering author "kids himself regularly," as he says.

> When I sit down to write a technical book I compute the number of manuscript pages I'll need. Each day, after I finish writing a few pages, I subtract these from the total needed. Thus, if I need 400 manuscript pages and write eight pages the first day, that leaves $400 - 8 = 392$ pages. Instead of frightening myself with the thought of writing a book I just concentrate on reducing the number of manuscript pages to zero. For when I've done this I have a book manuscript.

There are many other ways of approaching work that will make the job seem easier. Think of a particularly onerous task as a challenge to your patience. Don't think of getting the job done. Just concentrate on exerting your patience for a certain number of hours—the time it will take for you to do the work. Or set the job up in a number of stages. Reward yourself after completion of each stage. Don't work to get the job done—work instead for the reward due you at the end of each stage. While these techniques may seem unnecessary to some engineers and scientists, there are many men who find them useful.

Colin Carmichael, Editor of *Machine Design,* and a successful engineer, said,[1] "We talk about 'gain' of amplifying equipment—the boost given to an input signal. The engineer starts with an input which is the knowledge developed by previous generations plus the education to use and extend it. His gain is the boost he can give this input, to produce an output which advances the art and science of engineering." Any methods you can use to increase your output will be a big help to your career. Try fooling yourself now and then—you'll get more done with less effort.

Be Incautious at Times

Engineers and scientists propably exhibit more caution in their professional lives than any other well-educated group. This overcautious attitude can be traced to their training—they have been taught to question every fact until it can be verified. But an overcautious attitude can lead to a rut in your career from which it may be impossible for you to climb.

Try being incautious at times. Take a chance. Trust people for a change—instead of doubting their motives, their knowledge, and their abilities. The chances are good that most of these people will live up to your hopes. Some will astound you—their knowledge will far exceed your expectations. By dropping your overcautious attitude, you will make new friends. These new friends can open a completely different world to you—greater professional achievements, a new job, different hobbies, etc.

Take a chance in other areas of your life. If you've been debating a job change, study the pros and cons more carefully. You may make the change. Have you been studying the stock market, thinking of buying a vacation cottage, or wanting to take a special course in some subject? Be incautious, and do what you want to do, provided it isn't harmful to yourself or others. With a bit of incaution will come a new look at life. You will see that, in many phases of life, the gain is proportional to the risk.

Be alert for the opportunity to take a risk, for as Sophocles, the Greek dramatist, said, "Chance never helps those who do not help themselves." Once you learn to disregard caution at the proper time, you will begin to rise out of the specialist category into that of the generalist. And since the generalist is often the happiest and most successful person, you may find you achieve more when you take a chance.

Seek Privacy

You can often get more done when you are free of interruptions. So seek privacy when you work—be this at the office, on an airplane, on a train, or at home. Try getting to work an hour earlier. A busy research chemist says that he gets almost as much done between 8 and 9 A.M. as he does during the remainder of the day.

Avoid time-wasting conversations. Be jealous of your time and your privacy. "Solitude is as needful to the imagination as society is wholesome for the character," said James Russell Lowell. You may double or triple your output if you use the advantages of productive privacy.

Use Minutes and Seconds

Many engineers and scientists wail about the lack of time to achieve more. You may have listened to some of these men while they spent fifteen minutes telling you how busy they are. The truly busy engineer or scientist doesn't have time to talk about his full schedule. Instead, he works.

Try to save minutes—even seconds. Learn that saving fifteen minutes a day will give you 1.25 extra hours per week. In 50 weeks this is 62.5 hours, or almost two working weeks. If this time were devoted to useful professional activities, you could achieve much.

Remember that a little work each day can build to an impressive total. Thus, write 1,000 words a day, and in four days you have a 4,000-word technical or scientific paper. And in 100 days you have 100,000 words, the typical length of some technical and scientific books. Remember what Ben Franklin said, "Do not squander time, for that is the stuff life is made of."

Be Yourself

Some engineers and scientists have, in recent years, begun to adopt a pose of learned pomposity. To outsiders, and to most other engineers and scientists, such a pose is annoying and childish. So be yourself—if you're one of these posers. Don't try to imitate a college professor or a well-drilling roughneck. Instead, pinpoint your strongest skills and emphasize them.

Make the most of your past experience. If you lack certain types of experience—like public speaking, engineering design, scientific research, etc.—set out to acquire some in the desired areas. You can do this directly, by working in the area, or indirectly, by taking courses. Then develop the personality which seems to be most in keeping with your skills and experience. All three—skills, experience, and personality—will then blend to help make you a unique individual. By striving to be yourself you can develop into a unique person—wanted and respected in your professional and personal life. Should you ever falter in your attempt to be yourself, remember Emerson's famous words:

> There is a time in every man's education when he arrives at the conviction that envy is ignorance; that imitation is suicide; that he must take himself for better or worse as his portion; that though the wide universe is full of good, no kernel of nourishing corn can come to him but through his toil bestowed on that plot of ground which is given him to till. The power which resides in him is new in nature, and none but he knows what that is which he can do, nor does he know until he has tried.

Work, Think—Then Play

Do your job first, then take time off for play. Finishing your work can give you a feeling of relief. You'll enjoy your free time more after your job is done. Leaving a task before it is finished can spoil your free time because your unfinished work continues to haunt you.

Think before you play. Your free time will then be more productive. Many engineers, scientists, and mathematicians regularly give full thought to a problem for a certain period. Then they turn away from the problem, "forgetting it," as they say. Poincaré, the French mathematician who discovered Fuchsian functions, had many flashes of insight while at ease, *after* studying his problem. So think first, then relax.

Set New Goals

Once you begin to get things done faster, more efficiently, set new goals for yourself. Choose bigger, more difficult tasks. If your initial goal was to write a scientific paper, and you did, consider writing a series of papers. Too difficult? Remember, you resolved to get things done. You *can* write a series of papers, if you try. And after you've written the papers you can, if you really want to, write a scientific book.

Of course, your goals may be in other areas. Thus, you may want to achieve a high position in your professional society. Or you may wish to move ahead in your organization. In these and other goals your knowledge and skills in getting things done will be a big help to you. Rest awhile after you achieve one goal in getting things done. Then go on to newer, bigger, and more worthwhile goals. Keep trying and keep growing in your field, and your professional achievements will grow with you.

REFERENCE

[1] Colin Carmichael, "Measure of Success," *Machine Design,* Oct. 26, 1961.

HELPFUL READING

Josephs, Ray, *How to Gain an Extra Hour Every Day,* E. P. Dutton & Co., Inc., New York, 1955.

Laird, Donald A., *The Technique of Getting Things Done,* McGraw-Hill Book Company, Inc., New York, 1947.

Tuska, C. D., *Inventors and Inventions,* McGraw-Hill Book Company, Inc., New York, 1957.

Uris, Auren, *The Efficient Executive,* McGraw-Hill Book Company, Inc., New York, 1957.

———, *Developing Your Executive Skills,* McGraw-Hill Book Company, Inc., New York, 1955.

Wright, Milton, *Managing Yourself,* 2d ed., McGraw-Hill Book Company, Inc., New York, 1949.

PART 3

Broaden Your Professional Horizons

24

Obtain Your Professional License

> ... the principal factor in the lateness of the development of engineering registration licensing has not been so much that people are unaware of engineers, as that engineers are unaware of people, including themselves. That is to say that the typical engineer thinks of engineering, but does not think of engineers. . . .
>
> MELVIN NORD, P.E.[1]

Every engineer—and many a scientist—can benefit himself, his career, and his profession by obtaining a professional engineer's license. The advantages of being licensed are many and make the effort to obtain a license worthwhile. In this chapter you will learn the steps you can take to secure your license in the fastest and surest way.

Advantages of Being Licensed

Holding the professional engineer's license (1) gives you added prestige and recognition in your organization and community, (2) permits you to perform consulting engineering services in the state in which you are registered, (3) may entitle you to higher earnings in your organization, (4) helps you advance your profession by attesting to your abilities, (5) protects you and the profession from incompetent persons who might try to practice engineering, (6) establishes a professional consciousness in engineers and the public, and (7) encourages an upgrading of technical development, leading to a better life for all.

The professional engineer is recognized as an important member of society. Today, every state has a registration law governing the licensing of engineers who practice within its borders. As an engineer you owe it to yourself and your associates to fulfill the requirements for professional registration. No engineer who ever obtained his professional license ever regretted his action but many a man who was too lazy to seek and obtain his license has been sorry. Hanging your license in your office and keeping your seal in your desk drawer will never harm you. Not having the license when an important job comes up can make you bitter and self-critical. So if you don't have the license, start doing something now about getting it.

Check Yourself Now

Here are thirteen simple questions, with answers, that show you the necessary qualifications for the professional engineer's (P.E.) license. These are questions that are often asked by engineers and scientists everywhere. Try these questions now and see how well you rate.[1]

PROFESSIONAL ENGINEER'S LICENSE CHECKLIST

	YES	NO
1. Are you over 21 years of age?	( )	( )
2. Are you a U.S. citizen or have you declared your intention to become one?	( )	( )
3. Are you a high-school graduate or have you approved equivalent education?	( )	( )
4. Are you engaged in engineering work?	( )	( )
5. Do you hold an engineering degree?	( )	( )
6. Do you hold a science degree?	( )	( )
7. If you do not hold a degree, do you have a high-school diploma and 8 to 15 years engineering experience?	( )	( )

	YES	NO
8. Do you have an engineering degree and 4 years of engineering experience?	( )	( )
9. Do you have a science degree and 4 to 6 years of engineering experience?	( )	( )
10. Do you have an associate degree and 4 to 6 years of engineering experience?	( )	( )
11. Are you working in engineering during the day and enrolled in an accredited undergraduate evening engineering school and studying for a degree?	( )	( )
12. Do you hold a graduate engineering degree?	( )	( )
13. Are you a member of the Armed Services?	( )	( )

ANSWERS

1. You must be at least 25 years of age to qualify for the Professional Engineer's examination. However, you can sit for the Engineer-in-Training (E.I.T.) examination when you are 21 years of age.
2. You must be a citizen of the U.S. or have declared your intention to become a U.S. citizen before you can qualify for the P.E. license. However, a few states bordering on Canada may waive the U.S. citizenship requirement.
3. You must be a high-school graduate or have approved equivalent education to qualify.
4. You *must* be engaged in engineering work to qualify for the P.E. license.
5. You do *not* have to hold an engineering degree to qualify for the P.E. license, *provided* you have sufficient *approved* engineering experience. If you do have an engineering degree the total legal minimum approved experience required may be reduced.
6. With a science degree the total legal minimum approved experience required may be reduced, but not at as great a rate as with an engineering degree.
7. You may qualify for the P.E. license if you are a high-school graduate and have 8 to 15 years of approved engineering experience. While a degree is *not* required, study of engineering courses favors the applicant.
8. To qualify for the E.I.T. examination you must have at least 4 years of experience, if you hold a degree from a *non-accredited*

engineering school. Graduation from an *accredited* engineering school may qualify you for the E.I.T. examination immediately after graduation. Then, with 4 years of approved experience, you may qualify for the P.E. examination.

9. You must have 4 to 6 years of approved engineering experience, depending on the curricula pursued.
10. You must have 4 to 6 years of approved engineering experience, depending on the school attended and the curricula pursued.
11. Experience obtained *after* completion of the normal junior (sixth semester) year engineering curricula subjects will be considered and evaluated by the board of examiners and may be credited toward the approved experience required for the license.
12. Graduate study in engineering may be accepted in lieu of engineering experience. But examining boards usually limit the maximum experience credit to one year.
13. Engineering experience acquired in the armed services after graduation from engineering school is acceptable provided it is of a satisfactory grade.

Do You Know the Requirements?

To obtain a professional engineer's license, you must (1) meet the requirements of the state board as to age, education, citizenship, and experience, (2) pass a written or oral examination, and (3) have satisfactory character references. Let's study each of these items.

(1) Each state has specific requirements for registration as a professional engineer. These requirements cannot be precisely defined because of the differences in the registration laws of the individual states and territorial possessions. There are, however, seven general requirements,[1] the details of which may vary to some degree from state to state. These requirements are summarized in the checklist above and are: *age* (question 1); *citizenship* (question 2); *graduation* (question 3); *degree* (questions 5, 6, and 7); *experience* (questions 8, 9, and 10); *character* (usually five references attesting to the applicant's character and integrity are needed, three of which must be from licensed professional engineers); *examination* [see item (2) in the discussion below].

Some candidates have difficulty understanding the *experience* requirements. The basic objective of the requirements for qualifying experience is to make sure that the applicant has acquired, through actual practice in engineering, professional judgment, capacity, and competence in the application of engineering sciences. The quality of the experience should be such that it demonstrates that the applicant has developed technical skill and initiative in the correct application of engineering science. This experience must also show sound engineering judgment in the creative application of engineering principles and in the review of such application by others and the capacity to accept responsibility for engineering work of a professional character.

(2) Unless waived by the examining board, every applicant must pass all parts of the professional examination given by the board. Today the majority of persons obtaining a professional engineer's license are required to take and pass a written examination.

Most states give the examination in two parts. The first part is often called the Engineer-in-Training (E.I.T.) examination. It is also known as the *preliminary* or *first day's* examination. This examination is designed to test your facility in mathematics and engineering theory. The topics that may be covered in this examination include mathematics, engineering economics, physics, mechanics, electricity, magnetism, heat, light, sound, chemistry, electrical power, statics, dynamics, strength of materials, thermodynamics, and heat power.

In many states the E.I.T. examination can be taken immediately after the candidate obtains his undergraduate engineering degree. If at all possible the candidate should take the E.I.T. examination at that time because all these subjects are still fresh in his mind. In some states (New York and New Jersey) the E.I.T. examination is the open-book type. California has a combined open- and closed-book E.I.T. examination.

The second day's or professional examination requires you to show proper judgment in evaluating correct formulas, economic considerations, and practical approaches to the solution of engineering problems. The problem wording and terminology are arranged so you must call on your experience to develop an acceptable solution. An essay-type question or a critical description of a project with which you have some personal experience may be included. All questions in the professional examination involve the application of scientific principles to the solution of everyday engineering problems of the type you meet in the field or the office.

You are asked, on the second day's examination, to solve problems in your specialty—chemical, civil, mechanical, or electrical engineering. Besides the specialty questions most professional examinations include some questions on civil engineering. Thus, you might have one or more questions on fluid mechanics, materials of construction, surveying, etc. Here is a list of subjects given on the typical professional examination in electrical engineering; electric and magnetic circuits; basic and industrial electronics; electrical measurements; a-c and d-c machinery; telephone and telegraph communications; radio and television; power generation, transmission, and distribution; machine design; thermodynamics; fluid mechanics; materials of construction.

Steps in Obtaining Your License

There are fifteen basic steps in obtaining your professional engineer's license. These are: (1) Write your state board of registration requesting an application form and a copy of the law governing registration in the state in which you reside. (2) Study the law carefully to learn if you qualify for the examination you wish to take—E.I.T. or P.E. (3) Fill out the application form, being as neat and as complete as possible (seek guidance on the correct way to fill out the application if any parts of it puzzle you). (4) Send the application, together with any fee required, to the address specified by the state board. (5) Immediately begin to review the subjects on which you will be examined (you can attend review courses, take correspondence courses, or use special review texts). (6) As soon as you are notified of the time, date, and place of the examination, make arrangements to obtain time off from your job. (7) Pace your review studies so you'll have enough time to cover all the topics you will meet on the examination. (8) Get a full night's sleep prior to the examination. (9) Get to the examination room *ahead* of time. (10) Be calm and confident while taking the examination. (11) Solve the easiest problems first—this will build your confidence and help you attain a higher grade. (12) Allot yourself an equal amount of time to solve each problem—in a four-hour examination having eight problems, allow half an hour per problem. (13) If you solve some problems in less than the allotted time, go on immediately to the next problem—by working this way you will have more time available for solving the more difficult problems. (14) Once you finish the examination, forget it until you are notified of the outcome. (15)

If you fail, take the examination again, as soon as the state board allows you.

Prepare Well if You Wish to Pass

If you've been out of engineering school four years or more, the chances are good that you've forgotten much of the material you haven't been actively working with. To ensure a passing grade on the examination, you should review this material. There are several helpful ways to review subjects for the examination—refresher courses, correspondence courses, and special review texts. Refresher courses require that you appear for the class and solve the problems presented. If you're the kind of person who requires some urging to complete your studies, then the refresher course is for you. Seeing others around you complete their work will stimulate you to keep up with them. You will progress at the same rate as the class and will be prepared for the examination when it comes.

Correspondence courses are excellent for those men who must travel considerably or who live in small towns far from cities in which refresher courses are given. A correspondence course will keep you studying and can be as effective as a refresher course.

Special review texts are as useful as refresher and correspondence courses. Well-written review texts permit you to study at your own speed. When the answers to problems are given in the text, as they are in those listed at the end of this chapter, you can easily check the accuracy of your results. With a review text you can study in the privacy of your home, while you travel, or at any other time that you have a few spare moments. Try a review text—it may help you more than you realize.[2–7] Review texts, correspondence and refresher courses are available for study of the basic engineering sciences (fluid mechanics, thermodynamics, heat, mechanics, machine design, and electrical fundamentals), and your specialty (chemical, civil, electrical, or mechanical engineering). Other topics covered include engineering economics, ethics, and structural engineering.

Licensure Without Examination

If you've had fifteen or more years of experience in engineering, there is a possibility that you can obtain your professional license without undergoing a written examination. Many state boards recognize eminence—outstanding accomplishments in a particular field of engineering. Each state board of examiners applies its own rules on the meaning of eminence. Therefore it is impossible to predict exactly what criteria a particular state board will establish for a particular candidate.

Certain qualifications [1] may, however, be considered as evidence of eminence in the engineering profession. Check yourself now on the following qualifications and see how you rate.

1. *Demonstrated* adherence to high ethical standards
2. *Consistent* integrity in the practice of engineering
3. *Outstanding* professional ability, determined by studying your record for:
    a. Responsibility given you by your employer
    b. Publications in technical journals and books
    c. Work on professional society technical committees
    d. Patents and important developments and achievements
    e. Advanced degrees and studies
    f. Respect of fellow engineers
4. *Interest* in your profession
    a. Membership in professional, technical, and learned societies
    b. Work on society education committees—such as college- or high-school visiting committees
5. *Interest* in your community
    a. Boy Scouts or Sea Scouts of America
    b. Community chests
    c. Board of education or school board
    d. Parent-teacher associations
    e. American Red Cross
    f. Other recognized nonpolitical activities

Study the above list carefully. Note that it contains many items discussed in detail in this book—working well with your boss, writing technical material for publication, activities in your professional society, advanced studies, and community activities. If you rated your achievements fairly high in this list—say you've worked in ten of the sixteen categories listed—write your state board of examiners, asking them about licensure by eminence. Be sure that you have at least 15 years engineering experience before attempting to obtain a license by this means. Some states also set a minimum age for licensure by eminence. This age varies from a minimum of 35 years to 50 years, depending upon the state.

Don't become overhopeful about licensure by eminence. In general, it is becoming more and more difficult as states gain greater experience with license laws. But if you score high on the checklist given earlier and meet the other rules of your state, try the eminence route. The worst that can happen is that you'll be told you must take part or all of the written examination.

Being Licensed Is Important

Don't, like so many other engineers, shrug off the importance of being licensed. Having the professional license will *never* harm you. But not having the license can be a major roadblock to greater professional achievements. So begin today to take the steps that will lead you to the license. Once you are a professional engineer and can use the P.E. letters after your name, you will work with a greater sense of fulfillment for yourself and your profession.

REFERENCES

[1] John Constance, *How to Become a Professional Engineer,* McGraw-Hill Book Company, Inc., New York, 1958.

[2] Max Kurtz, *Engineering Economics for Professional Engineers' Examinations,* McGraw-Hill Book Company, Inc., New York, 1959.

[3] Max Kurtz, *Structural Engineering for Professional Engineers' Examinations,* McGraw-Hill Book Company, Inc., New York, 1961.

[4] William S. LaLonde, *Professional Engineers' Examination Questions and Answers,* 2d ed., McGraw-Hill Book Company, Inc., New York, 1960.

[5] Lloyd Polentz, *Engineering Fundamentals for Professional Engineers' Examinations,* McGraw-Hill Book Company, Inc., New York, 1961.

[6] John Constance, *Electrical Engineering for Professional Engineers' Examinations,* McGraw-Hill Book Company, Inc., New York, 1959.

[7] John Constance, *Mechanical Engineering for Professional Engineers' Examinations: Including Questions and Answers for Engineer-In-Training Review,* McGraw-Hill Book Company, Inc., New York, 1962.

25

Don't Be a Technical Laggard

> The man who utilizes his leisure by studying at home is usually increasing his ability. I can subscribe to that method because I studied that way and know its benefits.
>
> WALTER P. CHRYSLER

If you've been out of school eight to ten years, you must get back to your professional books if you plan to stay in engineering or science. Technology and science are advancing so rapidly today that ten years after graduation your education is far behind current practice. This is vividly shown in Dr. Thomas Stelson's remark quoted in Chapter 10 of this book.

Don't feel sorry for yourself because this condition exists. If it didn't, the opportunities and careers open to engineers and scientists would be severely limited. Welcome the advances of science and technology. These advances provide you with a greater chance to achieve more in your profession, to raise your income, and to acquire a greater number and higher quality of possessions for your family and yourself. Engineering and scientific advances will also provide a continual challenge to your understanding, creativity, and problem-solving ability. Such a challenge will keep your mind young, alert, and active. You will be better prepared to meet the changing conditions in your professional and personal life as time goes by.

Knowledge Expands

Our technical and scientific knowledge is doubling every ten to fifteen years, depending on whose forecasts you use. This expansion, and its effect on technical and scientific personnel, is recognized by many major firms. Thus, you'll find that Beckman Instruments, Varian Associates, Western Electric, General Dynamics, United Aircraft Corporation, Martin-Marietta Corporation, and hundreds of others assist their engineers and scientists in acquiring more knowledge through formal and informal courses of all kinds.

"But," you say, "I don't work for an organization that will help me pay the expense of a course or two. And have you noticed, friend, how the cost per credit has skyrocketed? How can I help from becoming a 'technical laggard,' as you've labeled me?"

It's easy to avoid becoming a technical laggard—no matter where you work, no matter what your specialty, no matter how small your income may be. In the remainder of this chapter you will learn of the many ways you can acquire new knowledge. At least one of the ways will be applicable in *your* circumstances—perhaps several will. Read on, and you will see how easy it is to avoid becoming a technical laggard.

Eight Methods

Successful engineers and scientists today have at least eight ways of keeping up with developments in fast-moving technologies and sciences. These are: (1) self-study, (2) correspondence courses, (3) extension courses, (4) journals and magazines, (5) professional society activities, (6) research leaves, (7) teaching, and (8) indexes and abstracts. Let's look at each method and see how you can apply it in your career.

Self-study pays dividends. All engineers and scientists study on their own. Every time you open a handbook, professional reference book, or good catalog, you must study the data presented before you can apply it. Why not make such study a regular habit, particularly in areas of new knowledge?

You can undertake self-study in many ways. One prominent mechanical engineer makes a practice of reading Marks' *Mechanical*

Engineers' Handbook for fifteen minutes each day. Since this handbook contains nearly 2,000 pages set in small type, it will take him more than a year to finish the book. Once he does he plans to switch to a handbook in a related field—electrical engineering. Over a four-year period this engineer will probably review every technical subject he studied as an undergraduate.

If the handbook technique doesn't interest you, then try the following: (*a*) Study a good, new professional reference book in your field every month. Force yourself to read the *entire* book. If there are certain portions of the book you can't follow—for example, advanced mathematics—get a book on this subject and study enough of the material to enable you to understand the math. (*b*) Regularly study a book in a new or rapidly developing field outside your own. Thus, if you're a chemist whose work involves some physics, study a series of books covering modern physics. (*c*) Develop a course of study in your own or a related field. Plan the course so it runs for a year or more, if you need that much time to cover the topics. Pattern the course after that given by a college or university you admire. Most catalogs provided by these institutions list the major topics the courses cover. Some catalogs even list the text used. (*d*) Plan a study program based on a group of catalogs in your own or in a related field. Many engineering and scientific catalogs contain a large amount of useful information. While catalog data may be frowned on by purists as being too empirical, it is to the catalog that every engineer or scientist must turn when he needs off-the-shelf items in his profession.

Force yourself to learn when you adopt a self-study program. Make use of every available moment—travel time, minutes you may spend in waiting rooms of any kind, even the time you spend waiting for your wife to dress. Moments spent this way can build to a sizable total, enabling you to increase your knowledge of new and related fields to a useful level. So begin today—start a self-study program that you can continue for the remainder of your career. You will contribute more to your profession and give yourself a fuller and happier life.

Try a correspondence course. Millions of people have broadened their knowledge through correspondence study. Many leading colleges and universities offer excellent correspondence courses in engineering and science today. Typical institutions include the University of Chicago, University of Wisconsin, North Carolina State College, University of Alabama, University of Arizona, University of California, University of Colorado, and University of Michigan. Many institutions grant credit toward a degree for certain courses completed by correspondence. You *cannot*, however, obtain a recognized degree completely by correspondence.

Use correspondence courses if you travel extensively, if you're stationed overseas, or if you want to have study supervised by an instructor. See Chapter 29 of this book for additional reasons for taking correspondence courses to learn a new or related subject.

Extension courses are popular. Probably more engineers and scientists use extension courses as a means of keeping up to date than all other formal methods combined. Extension courses given by colleges and universities usually carry full credit toward an advanced or undergraduate degree, whether the course is given on campus or off campus. Though accurate statistics are unavailable, probably more extension courses are given off the campus than on the campus. Thus, the University of Alabama gives courses at Huntsville, Alabama to engineers and scientists working at NASA's Redstone Arsenal. A variety of industrial firms in the electronics and aerospace activities have similar arrangements with nearby or distant institutions, depending on the location of firm and school. Some courses are given in the plant during or after working hours.

Enroll in at least one extension course per semester if your firm offers the opportunity. The few hours per week you devote to the course and assignments will be well repaid in extra knowledge and skill. Also, completion of a course will probably be counted as a credit toward promotion and a salary increase. Continue taking courses for credit until you have amassed enough credits for a graduate degree. Many firms offer the chance to obtain an advanced degree as an inducement in their hiring programs for engineers and scientists.

Where you must pay tuition yourself, give careful consideration to the advantages of extension courses. Even when your income is smaller than you wish, a few dollars spent on an extension course at a recognized university or college is an excellent investment in your professional future. Ben Franklin recognized this when he said, "If a man empties his purse into his head no one can take it from him."

And Karl G. Rahdert pointed out, "Development of a man's ability in his job, like other kinds of learning, is a continuous process. A man who abandons his efforts to keep up-to-date professionally becomes obsolete just as surely as facilities and equipment do."

Lastly, don't overlook the many noncredit extension courses. These offer an excellent opportunity for you to keep up to date by attending lectures given by outstanding faculty members. Noncredit courses that give you a good knowledge of a subject are worth every moment you spend on them.

Journals and magazines educate. Read good professional journals and magazines regularly, if you wish to be well-informed about your own and related fields. "In science, read, by preference, the newest works," wrote Bulwer-Lytton. Today the newest works of engineering and science first appear in the good journals and magazines serving a given field.

Develop a reading pattern in each field. Thus, as an electronics engineer you might choose to read *Electronics* magazine and the *Proceedings of the IEEE* for current information about the field. To these two basic sources you might add a number of specialized journals or magazines, if you feel you require more detailed information. You might also add one or two journals or magazines from foreign countries. And a summary magazine like *Science News Letter* might also help you to keep up to date in a number of broad topics.

No matter what your journal or magazine choice is, read regularly, read thoroughly, clip and file important items, and refer to these items at a later date. You can choose the finest engineering and scientific journals in the world, but they will do you no good unless you read them. So form the habit of reading every issue as soon as possible after it arrives. Clip the useful material, pass the publication on, file it, or have it bound in permanent form.

Learn at your professional society. Papers, panels, people—these are the three means through which you can acquire much useful technical knowledge in your professional society. Be sure you make full use of all because each offers a different method for you to participate in the advances in your field.

Papers presented by other members, and by nonmembers, often report the newest developments in a field. So make it a practice to review the titles of all papers related to your field. Obtain copies of the papers you think would be helpful. Read these papers as soon as you receive them; apply the information presented.

Preparing a paper for presentation before your professional society can also be an enlightening experience. See Chapter 9 for hints on preparing technical and scientific papers.

Panel discussions, increasingly popular in many professional societies, give you a firsthand chance to learn from the experience of others. Since panel discussions are verbal, you will often receive information a man would not report in written form. Also, there's usually more time for audience discussion of the topic. The give and take of the panel, along with pro and con audience discussion, can provide a highly interesting and informative presentation of new knowledge. Try attending a few panels and see for yourself.

People you meet in your professional society can provide much useful information on late developments. Some engineers and scientists claim they obtain as much useful information from people they meet at professional society meetings as they do from the papers presented. This, of course, will vary from one individual to another. Make a habit of meeting as many new people and old friends as you possibly can whenever you attend a professional meeting. You will find that your knowledge of and interest in your field will grow greater with every meeting you attend.

Research leaves help you grow. During the past several years, the General Electric Research Laboratory has experimented with a program of research leaves designed to give some of its outstanding scientists an opportunity to spend six months in postgraduate studies in the United States and abroad. Other research laboratories, engineering and scientific firms, have similar arrangements for postgraduate study. These research leaves resemble the sabbatical leaves granted by colleges and universities. At General Electric Company the leaves are granted to a relatively small number of outstanding scientists to take advantage of opportunities for advanced study pertinent to their field of research.

Dr. John S. Kasper, physical chemist in the Metallurgy and Ceramics Research Department of GE's Research Laboratory, reported [1] on a research leave he was granted. He spent six months as a visiting scientist at the H. H. Wills Physics Laboratory of the University of Bristol, England. Dr. Kasper reported:

> My position was simultaneously as a student, a teacher, and an independent research worker. I attended formal lectures and seminars on topics of fundamental physics. In the capacity of a teacher I gave a series of lectures in crystallography and structural programs to various members of the physics department. My main research activity, in collaboration with Professor F. Charles Frank, was the investigation of the nature of complex alloy structures. I also studied research activities elsewhere in England by visiting laboratories, attended meetings of scientific societies, and maintained personal contacts.
>
> I found the temporary return to academic life most pleasant and worthwhile. In this climate of high intellectual curiosity, it was easy for me to abandon temporarily some of my special interests and to concentrate on more general and basic aspects of physics, as well as to become acquainted with new sets of problems. It was one of my goals to achieve a broadening experience, and in that respect my leave was successful.
>
> I found much inspiration and insight into problems from my close association with Professor Frank. Together we completed two papers on alloy structures. My visits to other institutions were stimulating and informative. I already had good contacts with members of the Cavendish Laboratory at Cambridge, and my relations were made more firm while I was in England. In my visits and attendance at scientific meetings many new things were brought to my attention. The establishment of the many personal contacts with fellow scientists is something which I shall always cherish and which I believe is of direct value to the Research Laboratory as well. I personally have had a full, richly rewarding six months which I shall treasure for some time. It is also clear to me that other individuals can benefit equally from the research-leave policy. . . .

There you have the advantages one scientist obtained from a research leave. Should your organization ever offer you such a leave, take immediate action and accept the leave. For as Dr. Kasper says," It was one of my goals to achieve a broadening experience, and in that respect my leave was successful." Your leave will be successful too if you have the right goals.

Try teaching. You can increase your knowledge of a given subject by accepting an assignment to teach it in your organization or a local educational institution. Dr. Kasper, who taught while on the research leave described above, said, "It was gratifying to give my course of lectures on crystallography because, while I was presumably instructing, I learned a good deal myself. This resulted not only from the required organization of material and review of recent advances, but also from the mature and challenging questions that were asked. I was led to new viewpoints in my own areas of specialization."

Dr. Kasper's report on his teaching experience should encourage you to actively seek an assignment to teach some subject. Not only do you learn, you also help others. Further, the teaching experience will be a credit on your employment record and may be an important factor in your next promotion. Also, you may be paid for the time you spend teaching. Lastly, don't overlook the chance to convert your course notes into a good textbook. For example, Douglas Sailor and Frank Riley, both engineers at Lockheed's Missile and Space Division, gave a series of lectures on space science at their local high schools. Their lecture notes were later converted into a successful technical book, *Space Systems Engineering*. See Chapter 9 for hints on writing.

Learn from indexes and abstracts. Make a regular study of the indexes and abstracts devoted to your main fields of interest. Visit your company or local library at least once a week to study the new literature in your field. By using indexes like the *Engineering Index* or abstracts like the *Chemical Abstracts,* you can save much time in acquiring useful information. Once you find an item of interest, read the original text. You may miss much useful information if you do not read the original text of the item.

Study for an Advanced Degree

Probably more engineers and scientists are studying for advanced degrees today than ever before in our history. Why? Because technology and science are increasingly requiring greater skills and more advanced knowledge of mathematics, physics, chemistry, and hundreds of other subjects. Twenty-five years ago the engineer or scientist with a bachelor's degree could look forward to a lifetime career without much further study. Today the engineer or scientist holding only the bachelor's degree is at a definite disadvantage. To achieve more today you must know more; to know more you must study more. Every industry recognizes these facts.

Hundreds of firms in a variety of fields now offer their engineers and scientists a chance to obtain the graduate degree in a technical or scientific specialty. Many current educational programs are a blend of time-off and tuition allowances. For example, Beckman Instruments [2] encouraged its engineers to obtain advanced degrees by allowing them to work a thirty-hour week while collecting full pay and tuition refunds. Varian Associates evolved a Stanford-Varian Honors Cooperative Program under which qualified engineers earn a master's degree in two years at Stanford while they work at Varian. Western Electric's Graduate Engineering Training Program selects experienced engineers from time to time to take two- to four-week courses in special subjects. Many of General Dynamics Corporation's far-flung enterprises are backstopped by local educational institutions that provide graduate training for company engineers. And United Aircraft Corporation has been instrumental in setting up an educational center known as the Hartford Graduate Engineering Center. The center, a branch of Rensselaer Polytechnic Institute, offers students an M.S. in a variety of curricula including aeronautical engineering, applied mechanics, data processing, metallurgy, mechanical engineering, management, automatic control, electrical engineering, nuclear science, and nuclear engineering.

Many other organizations offer graduate training of some kind. Take advantage of every course that will help you. Obtaining the graduate degree will update your knowledge and skills, increase your earnings, and put you in a better position to achieve more in your profession.

Some engineers have a habit of comparing themselves with doctors and lawyers so far as status, income, and professional recognition go. The conclusion of these engineers invariably is that they do not receive the same recognition, income, and freedom that doctors and lawyers do. If you're in the habit of making this comparison, think for a moment. The medical doctor holds the equivalent of a doctorate degree in engineering or science—so do many lawyers. How can you as a holder of a bachelor's degree expect the same recognition as a holder of a doctorate? In medicine the holder of a bachelor's degree is confined to technician-level duties like nursing, laboratory management, etc. If you aspire to rise higher, to achieve more, you'll have a much better chance if you hold an advanced degree. Why not begin working toward that degree today? Then you'll have fewer reasons to complain about poor treatment by your organization. Remember: Lost time is never found again.

Dr. D. H. Ewing, while vice president of Research and Engineering at Radio Corporation of America said:

> Keeping our technical skills at a high enough level has become an increasingly difficult job with the great outpouring of devices and systems—principally electronic—whose complexity is beyond the comprehension of the average citizen. . . . In our mechandising we talk a lot these days about "planned obsolescence." In the case of our know-how, we seem to be suffering from "unplanned obsolescence." . . . It is a major responsibility of all men and women with technical training, from the mechanic to the theoretical physicist, to restore the tradition of know-how to this country.[3]

Don't Neglect Nontechnical Studies

So far in this chapter you've been learning of ways to improve your technical and scientific know-how. But this isn't the only area in which you should strive to keep up to date. As Clinton J. Chamberlain [4] wrote in the *Harvard Business Review*, "Intensive exposure to formal logic and philosophy could well be of far more importance to a computer designer than another course in transistor circuitry."

Studies show that an engineer or scientist with a well-balanced education which blends technology with the humanities is likely to achieve more than the narrow specialist. Further, the man with the broader education will probably be a happier and a better adjusted person. So begin now to broaden your knowledge of subjects outside your professional field. Consider subjects like art, music, philosophy, psychology, etc. You can use all the techniques given earlier in this chapter. An easy way to begin is to read better literature when you travel. Instead of picking up a cheap detective novel, try Conrad, Shakespeare, Milton, Dewey, or other writers of this calibre. You'll not only enjoy what you read, you'll also begin to move out of the intellectual rut that traps so many technical laggards.

Two of America's best-known modern writers had wise words of advice for every engineer and scientist interested in improving himself. Mark Twain said, "Keep away from people who try to belittle your ambitions. Small people always do that, but the really great make you feel that you, too, can become great." And Eugene O'Neill wrote, "A man's work is in danger of deteriorating when he thinks he has found the one best formula for doing it. If he thinks that, he is likely to feel that all he needs is merely to go on repeating himself . . . so long as a person is searching for better ways of doing things his work is fairly safe."

REFERENCES

[1] "An Engineer Takes a Sabbatical," *Product Engineering,* Mar. 31, 1958.

[2] "How the Engineer Can Go Back to School," *Product Engineering,* Mar. 31, 1958.

[3] D. H. Ewing, "Unplanned Obsolescence," *Dun's Review and Modern Industry,* May, 1959.

[4] Clinton J. Chamberlain, "Coming Era in Engineering Management," *Harvard Business Review,* September–October, 1961.

HELPFUL READING

Bradshaw, F. F., "The Citizen Engineer," *Mechanical Engineering,* June, 1962.

Carmichael, Colin, "Over the Hill?" *Machine Design,* Mar. 29, 1962.

Cunningham, W. A., "Professionalism is a Two-Mile Road," *Chemical Engineering,* June 26, 1961.

Devries, H. B., "Have You Read . . . ?" *Mechanical Engineering,* February, 1962.

Drucker, Peter, "Education in the New Technology," *Think,* June, 1962.

Foster, M. B., "What It Takes to Be an Executive," *Petroleum Refiner,* November, 1957.

Fremed, R. F., "What Basis for Comparing Salaries?" *Chemical Engineering,* Oct. 3, 1960.

Kempfer, E. D., *Home Study Blue Book,* National Home Study Council, Washington, D.C., 1958.

Radusepp, Eugene, "What Management Expects of Engineers," *Machine Design,* Apr. 26, 1962.

Sizelove, O. J., "Developing from an Engineer to an Administrator," *Mechanical Engineering,* April, 1959.

de Solla Price, D. J., "The Acceleration of Science—Crisis in Our Technological Civilization," *Product Engineering,* March 6, 1961.

Tangerman, Elmer, "The Engineer Comes into His Own," *Product Engineering,* Mar. 31, 1958.

———, "Learning Is Power," *Product Engineering,* June 30, 1958.

"Ten Years Out? Go Back to the Books," *Chemical Engineering,* May 29, 1961.

Tyler, Chaplin, "Steps in Becoming a Better Manager," *Chemical Engineering,* Apr. 30, 1962.

Van Lennep, D. J., "Why Some Succeed and Others Fail," *The Executive,* August, 1962.

26

Contribute More Through Your Professional Society

> He who waits to do a great deal of good at once, will never do anything.
>
> SAMUEL JOHNSON

Almost every engineer and scientist holds membership in at least one professional society. Some men are members of eight or ten societies and associations. Much depends on a man's inclination, his interests, and the time he can devote to professional society activities. You needn't belong to a large number of societies to achieve more in your profession. But you should belong to at least one socitey—the one that represents your major field of specialty. Thus, an electronics engineer might belong to the Institute of Electrical and Electronic Engineers, a rocket engineer to the American Rocket Society,* etc. One society, served well, can do more for you and you for it, than eight societies whose meetings you never attend.

Professional Societies Are Important

Your society serves you in many ways. It is the voice of your profession, informing everyone about the aims, activities, progress, and

* Now named Institute of Aeronautics and Astronautics.

problems of the profession. Your society may draft technical standards, prepare a code of ethics, publish learned papers authored by its members and nonmembers, serve as an employment counselor, hold technical or scientific meetings, finance and staff a specialized library, and engage in hundreds of other activities. Today some societies and associations even have group hospitalization and insurance plans for their members.

Do you use your society as a means to contribute more to your profession? Many engineers and scientists do not. Instead, they leave the work of the society to a few ancient hands who have been around for more years than anyone can remember. As a result the society stumbles along, failing to render the service it should. But since a professional society is organized to serve its members, who can be blamed if it fails to accomplish this task? Only the members because it is they who infuse knowledge, skill, and energy into a society.

Every professional society in existence today needs more members who participate in the activities of the society. Regardless of your educational background and technical interest—engineering or scientific—you can contribute more to your society and advance your own career further by active participation in your society. Let's see how.

Contribute to Your Field

Any man who spends thirty to forty years in a profession can make a valuable contribution to his field. How? He can do it (1) by presenting a learned paper before his professional society on some phase of his work; (2) by serving on professional committees organized by the society; (3) by taking action of some kind (voting, speaking, campaigning) for an important professional matter (licensing, ethics, laws, etc.); (4) by collecting funds for a society building campaign, charity drive, or other worthwhile cause; (5) by training younger men in important aspects of the profession; (6) by serving as a consultant within or outside the society on matters of importance to the society or the community; (7) by devoting a certain portion of his time to activities the society designates as important.

Contributing to your field may seem like an empty phrase. Yet it is probably the most important motivation you can have. For in contributing to your field, you can improve the lot of mankind and expand your career and knowledge. You never lose when you make a genuine contribution. As a further return you derive a satisfaction that cannot be measured in dollars but is a continuing source of pleasure throughout your life.

Present papers. Every professional society in the world today wants more technical and scientific papers of better quality. You thus have a better chance than ever before of presenting and publishing a learned paper. And in recent years most professional and scientific societies have altered their concept of what a good paper must contain. Today you'll find that many technical and scientific papers are (1) shorter, (2) less mathematical, (3) less historical, (4) more conscious of the human aspects of the profession, and (5) more informal in their language and handling of certain subjects.

What can you write a paper about? You can write about almost any subject of interest and importance to a few or many members of your professional society. Because a professional society is organized to serve its members, your paper will be given serious consideration even if it covers a subject of interest to only a small portion of the membership. Societies often publish papers that could not be handled by a business or trade magazine or paper because of limited interest in the subject.

Where do you get ideas for learned papers? Look at your daily activities. Ideas are everywhere. All you need do to latch on to a few is to train yourself to think in the right terms. As an engineer or scientist you are particularly fortunate. For you can find so many ideas for learned papers that you'll never have time to put them all to use. The key to effective papers is choice of the best ideas available to you.

Typical subjects for technical and scientific papers include: (*a*) results of a study or tests of a specific phenomena; (*b*) design procedures useful in planning a process, product, or device; (*c*) operating and maintenance histories of a machine, plant, or vehicle; (*d*) new practices in management, recruitment, research, and development; (*e*) data on expected performance, use, and reliability of equipment to accomplish a desired function. Many technical and scientific papers written today give a detailed discussion of a specific and limited problem. Papers covering a broad area of a subject may delve into historical aspects and may also present an extensive bibliography of the subject.

How papers are developed. Learned papers are either *solicited* or *contributed.* With a solicited paper you are asked by the society, or one of its divisions or committees, to prepare a paper on a subject with which you are familiar. Never turn down such a request. To be asked to prepare a paper is an honor. You were chosen after thorough investigation of your knowledge, skill, and ability.

With a contributed paper you choose a subject with which you are familiar and inform the society of your intention to prepare the paper. The professional society will either accept or decline your idea. Much depends on whether any papers on the same subject are being solicited, the number of papers available for publication, and other factors. You need not be a member of a society to write a paper for it. But most societies are more sympathetic to papers submitted by members than by nonmembers.

Writing the technical paper. To write a good technical paper, (1) prepare an outline, (2) gather data, (3) select your illustrations, (4) write the rough draft, (5) polish the rough draft, (6) check all your facts, (7) have the paper typed according to the society's specifications.

Use a simple style when writing the paper. Don't be wordy or pompous. Choose simple terms and expressions to convey information to your readers. Define all new words, terms, and symbols. Keep your sentences, thoughts, and definitions concise. Write for the typical member of the society—then your paper can be read and understood by most of the members.

Limit the paper to the maximum length recommended by the society. For example, the American Society of Mechanical Engineers suggests that the text not exceed 4,000 words. Use illustrations and tables where they help clarify your text and where they demonstrate results. Secure the necessary approvals and clearances before submitting the paper to the society. Prepare a 100-word abstract of the paper. State only the major findings, results, or conclusions in the abstract. Use a specific, explicit, and short title for the paper.

Processing of papers. Figure 26.1 shows the steps in the submission to and publication of a paper by the ASME. Many other societies follow a similar procedure. Study this flow diagram for it shows what you can expect to happen to your paper after it reaches

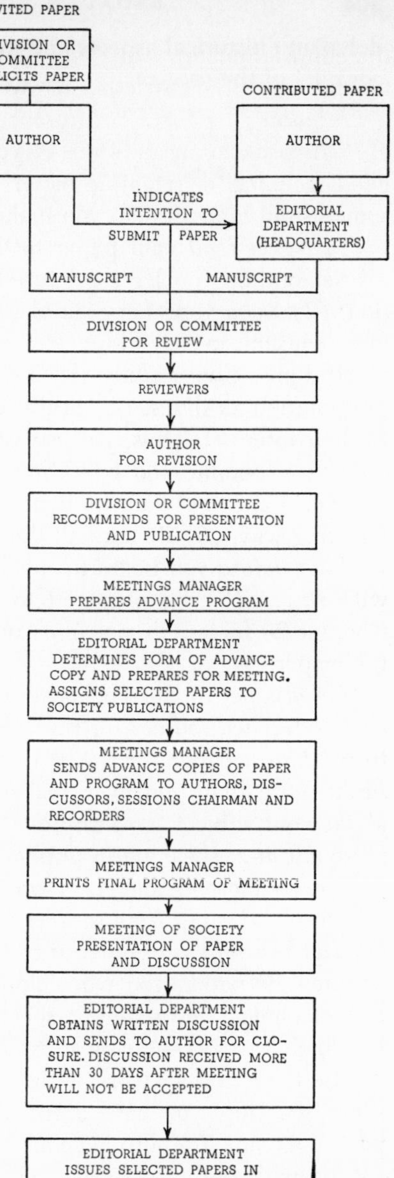

FIG. 26.1. Steps in submission to and publication of a technical paper by the American Society of Mechanical Engineers.

the editorial department. Note that your paper will be reviewed for technical accuracy, whether it was solicited or contributed.

Presenting a paper. To present a paper effectively, you must be able to speak to an audience of your fellow members. Don't ruin a good technical or scientific paper by a poor presentation. Here are some useful hints to help you make a better presentation.

(1) Never *read* your paper to the audience. Instead, speak from a brief outline you have prepared specifically for verbal presentation. (2) Keep your story simple—state your main findings, results, or conclusions in easy-to-understand language. (3) Use good, simple illustrations. Choose only the key illustrations from the paper for presentation as slides. (4) Study your audience and gear your talk to their interests. Thus, you can stress design data for younger audiences, economic and operating data for more mature audiences. (5) Don't *speak at* your audience—*talk with them* in a friendly, interesting way. (6) Be sure all the facilities in the meeting room are working *before* you begin to speak. (7) Watch out for booby traps with slides. (8) Be prepared to answer audience discussion. See Chapter 20 for additional hints on presenting a technical or scientific paper.

Plan a paper today. Technical and scientific papers constitute the main reference sources for researchers in every field. So plan today to prepare a paper for the next meeting of your society. You will profit more than you ever realized. For you will (*a*) learn more about your subject, (*b*) improve your writing ability, (*c*) become more adroit at speaking in public, and (*d*) make a lasting contribution to your profession. After presenting your paper you may have career job offerings from sources you never thought possible.

Don't overlook professional committees. In presenting a technical or scientific paper, you work alone, or with a few coauthors. Most large technical and scientific societies also offer you another sphere in which to work—the professional committee. On a professional committee you may work with as many as fifty other engineers or scientists. Here you have a chance to develop skills in working as part of a team. You also have an opportunity to meet and closely associate with the outstanding men in a given specialty. Your career and achievement goals widen, and a whole new area in which to make a contribution opens to you.

How can you serve on such a committee? There are three steps:

(1) Determine which committee in your society is closest to your technical or scientific interests. (2) Get to know one or more members of the committee. (3) Volunteer to help the committee in some of its activities. Once you prove you can work harmoniously with one or more committee members, you will almost certainly be appointed to the committee when a vacancy exists.

Look for a moment at the opportunities just one society offers to members interested in committee activities. The American Rocket Society * has nineteen technical committees devoted to the following specialties: astrodynamics, communications and instrumentation, electric propulsion, guidance and control, human factors and bioastronautics, hypersonics, liquid rockets, magnetohydrodynamics, missiles and space vehicles, nuclear propulsion, physics of the atmosphere and space, power systems, propellants and combustion, ramjets, solid propellant rockets, space law and sociology, structures and materials, test operations and support, and underwater propulsion. These technical committees, and subcommittees they may appoint, cover every modern phase of space technology. Members of these committees come from all fields—industry, government, research institutes, universities, and research laboratories.

Some technical and scientific society committees are organized in the simplest way, having a chairman, vice-chairman, and members. Other societies use a more complex structure. Thus, the Oil and Gas Power Division of ASME has an Executive Committee of five men. This committee is directed by the Chairman of the Division, who is also a member of and Chairman of the Executive Committee. Assisting the Executive Committee are twenty to thirty associates. These Associates serve on a variety of subcommittees, like Honors and Awards, Publicity, Publications, etc. Each year an Associate is appointed to the Executive Committee as one man retires from this committee. Being appointed to the Executive Committee means that you will be Chairman of the Division for one year. The chairmanship is an outstanding honor sought by many men.

Consider joining a technical committee of your society today. Being a member of a committee will give you a stronger professional voice than you could ever muster on your own. You will actively work with some of the outstanding members of your profession because they too will be serving on the committee. Your knowledge of

* Now named Institute of Aeronautics and Astronautics.

new developments will be far broader and more current. Committee discussions and decisions will stimulate your creative abilities. Because most committees review technical or scientific papers to be presented before the society, you will have a chance to study the work-in-progress of various authors. Examining and criticizing the work of others will teach you how to avoid certain mistakes in your own writing. Committees also choose the theme and topics for specific meetings of the society. As a member of a technical committee you'll have the opportunity to recommend paper subjects and meeting themes in accord with your own thoughts.

Many societies run annual exhibits in conjunction with their technical meetings. As a committee member you may have a chance to help in the exhibit business activities. These duties will teach you the economics of running an exhibit—and the problems to be expected.

Chances to honor well-known men in your profession will come your way because many technical committees select and screen the men to be honored. You will help and advise students who wish to specialize in your profession after graduation from school. Chairmen and vice-chairmen for technical sessions at meetings of the society are chosen from among committee members. Thus, you can look forward to wielding the symbolic gavel, indicating you are in charge of the meeting. Lastly, if you work well on the committee, you will probably receive a number of job offers from fellow committeemen. You will be pleasantly surprised to find that many of these job offers are well worth studying.

Collect funds; serve useful causes. Your society, sooner or later, will need extra funds for a specific use—a new headquarters building, a larger library, a special charity. Volunteer to help the next time your society asks for aid. Contribute some cash—try to give enough to make yourself feel a slight pinch of funds for two or three weeks. Such a gift will, of course, be welcomed by the society. And you will have a gratified feeling, knowing you have given to the point where it hurt.

After you've contributed money, volunteer your time and energy. Collecting funds for any society is a many-sided experience—frustrating, gratifying, annoying, perplexing. Try to collect funds from ten or more members of a society, and you'll learn more about human nature, inactive society members, and group participation than you ever thought possible. Your experiences while collecting funds may even sway you to decide that your society needs more voluntary helpers. If so, try to persuade other society members to join you in your voluntary tasks. As an example of what can be done by volunteer groups, the greater part of the $12-million cost of the twenty-story United Engineering Center building was raised by outstanding engineers, industrialists, scientists, and educators who collected and donated funds on a voluntary basis.

Voluntary work for your professional society can lead to worthwhile duties outside the society. Thus, James L. O'Neill, while chairman of the ASME Region II Civic Affairs Committee, was appointed chairman of the Mid-Hudson Science Advisory Council. This Council works with fifty-five high schools in five counties, arranging seminars and panel discussions and preparing workshop programs for both students and teachers. It assists at science fairs, arranges plant visits for students, and helps schools get surplus tools and supplies from industry. Mr. O'Neill is also chairman of the Advisory Board for Science Education, Board of Education, Poughkeepsie City School District. This board directs all secondary-school activities, was instrumental in adding Russian and German to the curriculum, sponsors an annual science fair and a sixteen-week exceptional-student science seminar. In addition to all these activities, Mr. O'Neill is the manager of the Industrial Sales Division at Daystrom Electric Corporation.

Attend society meetings. The main function of every technical and scientific society is the exchange of useful information among members. While you can learn much from studying published technical papers, you will usually learn more if you attend the meetings at which the papers are presented. In attending society meetings, both national and regional, you (1) learn new technical and scientific facts from the papers presented and people there; (2) meet people who are good for you—other engineers and scientists, educators, consultants, etc.; (3) have an opportunity to exchange ideas, methods, and techniques with the people you meet; and (4) can tour interesting plants, laboratories, and other installations where you can acquire useful information and data.

All the activities at the usual technical or scientific meeting will help broaden your knowledge of your field. And just as important, attendance at these meetings will help you become acquainted with more people in your profession. The larger the number of people you know, the greater your chances of success in a given field. So you never lose when you attend professional society meetings. What's more, the expense of attending meetings is almost always paid by your firm or employer.

Use society facilities. Almost every major technical and scientific society offers you many facilities in the national and regional headquarters. These facilities include meeting rooms with projection and amplifying equipment, a technical or scientific library, specialized publications in your area of interest, an employment agency, lounges, and conference rooms. Make use of these facilities whenever you need them. As a society member you are entitled to full use of the facilities offered to members.

Some societies provide personalized services for members requesting them. These services include receiving and recording telephone calls, office desk space, reproduction facilities, dictating equipment, receiving mail, and transcribing facilities. Such services are extremely useful while you are in the headquarters city, away from your usual sources for these services. Some engineers and scientists even use the society headquarters as a base of operations when looking for a new position. You're entitled by the terms of your membership to many privileges. Start using them today.

Societies and retirement. To older engineers and scientists professional societies offer as many advantages as they do to younger men. For every professional society will welcome the breadth and length of your experience in your chosen field. Thus, you'll find that you can prepare papers, serve on committees, and do voluntary work to the full extent of your abilities and time. Ted Robie, who was chief engineer of Fairbanks, Morse & Company for many years, became secretary of the Oil and Gas Power Division of ASME on his retirement from his engineering position. Ted likes his duties as secretary because they keep him in touch with the field he served for so many years. Since he attends all the meetings of the division, he periodically renews his friendships with many of the prominent men in the industry. Attendance at the meetings also keeps him abreast of the latest developments in the field.

Some professional societies also offer chances for part-time employment. Thus, you might find your talents sought through the society by firms needing consultants for special projects. Legal groups requiring expert witnesses in a given area may also be recommended to you by the society. Certain technical and scientific committees will pay for your services if you assist them with specific projects. Other opportunities for part-time employment by societies exist in a variety of areas like preparation of reports, writing of society-sponsored books, assisting in the planning and conducting of technical meetings, editing of papers, lecturing student groups, etc.

Don't overlook these and other chances offered by societies to extend your career into the later years of your life. The services you render to the society and mankind will fill you with a lasting satisfaction. As one retired physicist said:

> It wasn't until I reached seventy years of age that I really appreciated the full value of my scientific education. Contributing some time, energy, and knowledge to the Society created more joy in my life than I can describe. I've traveled all over the world for the society and have seen, for the first time, the valuable benefits mankind can derive from engineers and scientists working together to create a better life for all.

Get consulting work. As is shown in Chapter 30, technical and scientific societies offer you many chances to engage in professional consulting. By presenting several good papers in your specialty and serving on a committee devoted to your field, you can benefit handsomely. See Chapter 30 for details of how part- and full-time consultants use the opportunities offered by their professional society.

Take part in social activities. Many professional societies hold banquets, dances, dinners, and luncheons as part of their meeting programs. Others have a ladies auxiliary. These activities give you an opportunity to include your wife in some of your professional meetings. Take her along whenever you can. She'll enjoy getting away from her regular duties to meet your friends and associates. Knowing these people will assist her in discussing future activities with you.

If your society has a club nearby—like the engineer's club, chemists' club, etc.—consider joining it. Membership in such a club will broaden your friendships, give you a place to meet people, and provide a pleasant atmosphere for lunch or dinner.

HELPFUL READING

Hegarty, E. J., *How to Run Better Meetings*, McGraw-Hill Book Company, Inc., New York, 1957.

Robert, H. M., *Rules of Order Revised*, Scott, Foresman and Company, Chicago, 1951.

Sturgis, A. F., *Sturgis Standard Code of Parliamentary Procedure*, McGraw-Hill Book Company, Inc., New York, 1950.

27

Learn to Work with Specialists

Knowledge is power.

FRANCIS BACON

The great end of life is not knowledge but action.

THOMAS HUXLEY

As an engineer or scientist moving up to greater achievements you probably know as much or more about the mind of the technical specialist as anyone. There's just one hitch—some engineers and scientists forget many of their pet complaints when they become managers. As a result their attempts at supervising others are inept. Since you will almost always deal with specialists of some kind—technicians, engineers, scientists, etc., knowing how they think and what they seek from their careers can be a big asset to you on your way up.

What Technical Specialists Want

Many studies have been made of engineers and scientists to determine what these men want from their careers. Briefly, most want: [1]

- To pursue technical or scientific knowledge for its own sake
- Recognition from their peers for their engineering or scientific contributions
- Dollar rewards for their efforts

Of these three desires the first two cause the most trouble in the average business organization. Pursuing knowledge for its own sake does not, in general, produce a profit. And it is profit that businessmen and managers seek. Until the average engineer or scientist understands this and willingly cooperates to produce profit, both he and his manager (usually you) will have trouble.

Most organizations realize the importance to an engineer of recognition from his peers for contributions to his field. The same is true for scientists. So when an engineer or scientist prepares a professional paper, he is almost always permitted to travel at his organization's expense to a meeting at which the paper will be presented. A difficulty that sometimes arises in this situation is that management would prefer to have the engineer or scientist prepare the paper on his own time instead of business time. This requirement is distasteful to many technical men who wish to work on the paper at their place of employment. Arguments, misunderstandings, and bitterness often result.

Almost every engineer and scientist on a payroll today thinks he's underpaid. You'll find that some of these men, particularly those in their late thirties and early forties will malign management unmercifully when salary scales are discussed. Usually, though, the men who complain most are those who are willing to do the least to advance themselves. An occasional exception will, however, occur and a deserving, ambitious man will be paid less than he's entitled to earn.

How Technical Specialists View Their Work

Many surveys [1] indicate that most engineers and scientists are dissatisfied with their jobs. These men believe they:

- Lack chances for advancement
- Have little voice in controlling their organization
- Are underpaid

The same surveys show that many engineers and scientists do not hold management in high esteem. These men believe that managements:

- Are too dollar-minded
- Have little depth of knowledge
- Oversimplify problems
- Manipulate people for their own purposes

If you give these beliefs some thought, you will recognize many of them as common complaints of working engineers and scientists. In fact [1] a "study of the Public Opinion Index reveals great strain between the scientific intellectuals and management. Job satisfaction of these people, as gauged by standard morale measures, is consistently below that reported for management, foremen and other white collar groups. The conflict, it would appear, is over status and values."

Knowing these and other attitudes of engineers and scientists can help you work more smoothly and efficiently with these men. For you will find that as you advance in responsibility, you will have more problems with people than with any other aspect of your work.

Build the profit objective. Regardless of how knowledge-oriented an engineer or scientist may be, he still has to feed, house, and clothe his family and provide for the education of his children. If he is an active person, he'll have a hobby, which will probably require that he spend some money on it. Thus, though an engineer or scientist loves to pursue knowledge for its own sake, he still has a number of responsibilities to himself and others which must be met. In a profit-making organization of any kind the engineer or scientist can increase his income if he contributes more to the profit potential of his firm.

How can you, dealing with these engineering and scientific specialists, get them to recognize the importance of contributing to the profit potential of their organization? It isn't easy, but you can do it by following these steps:

1. Impress on each specialist, at every opportunity, the importance of his work to the organization's profit potential. Whenever possible, show him how a task he has performed has generated income for the organization. Keep hammering away at the profit theme. As one

project engineer put it, "I spent a year and a half telling our design engineers and draftsmen how important their work was to our profit. It took about that long for the story to sink in. Now they're telling me how important it is to be profit-conscious. Today our group has much more enthusiasm and efficiency. But it took time." So don't expect to convert knowledge-seeking engineers and scientists to profit-oriented businessmen in a month. It's a slow, and sometimes painful, conversion.

2. Show each man how *his* earnings are related to the organization's profit. Money is one of the strongest motivators known. And no matter how dedicated to knowledge an engineer or scientist may be, his income is still a matter of extreme importance to him. When shown that his income is usually a direct function of his contribution, the average engineer or scientist will usually exert an extreme effort to increase his output. For, as the ancients observed, "Money answereth all things."

3. Emphasize the fundamentals of profits. In the Western world the capitalistic system, based on profit making, provides the highest standard of living known. Profits are not, as some intellectuals contend, "dirty." Instead, profits are the legitimate returns for superior business judgment and the successful overcoming of risks. So the next time an engineer or scientist scoffs or derides profits while complaining of how poorly rewarded he is, stop him and (*a*) point out that profits are a legitimate return for risks taken and judgment exercised, and (*b*) that they are as important to *him* as to his organization. For when the organization profits, so does the engineer or scientist.

4. Develop skill in dealing with the many false attitudes assumed by some engineers and scientists. Thus, you'll find some men assuming a bored attitude. They're really not bored—but by appearing to be they acquire a feeling of superiority. Such men can be superior—if they make more than a normal contribution to their field. The best way to cut through this false attitude is to shock the man into realizing that poses don't pay off—contributions do.

Another common attitude is that of extreme complexity—"sophisticated" is the term many engineers like to use. While it is true that our entire technology is becoming more complex, to hide behind a facade of differential equations and Boolean algebra does little to increase understanding in an organization. The "sophisticated" engineer or scientist typically regards management personnel as overrewarded boobs. Instead of trying to understand why management personnel are better rewarded, the "sophisticated" technical man ducks behind his belief that the equations on his scratch pad make him a better man than his manager. Actually, neither is better nor superior. Each makes his contribution in his chosen area. The manager would be lost without the contributions of the specialist. The specialist, on the other hand, would have extreme difficulty marketing his contributions without the know-how of the manager. You can break down some of the wall of sophistication by bringing management and specialists together at meetings designed to get them to air their common problems. Properly run meetings (Chapter 22) can do much to relieve tension and develop mutual respect and understanding.

There are many other attitudes you'll have to learn to cope with. These include (*a*) the know-it-all who won't give you a chance to finish a sentence before interrupting, (*b*) the overly humble technical specialist who uses his humility to garner special favors, (*c*) the helpless individual who needs help with and complete supervision of every task assigned him, and (*d*) the complainer who runs to you with every minor injustice or slight. In coping with any of these attitudes, do not hesitate to be firm or even tough. Sometimes the only way to make a man realize he's being childish is by showing him the smallness of his attitude. This will often require that you be firm or tough. But never become enamored of an image of toughness for yourself because you will then be adopting an attitude which is as harmful as those just described. Remember that you, too, were once an engineer or scientist and that you probably had problems understanding profit objectives and management actions. The supervisor who was adroit at human relations (Chapter 16) probably helped you considerably. So apply intelligent human relations whenever possible in place of tough authority.

5. Listen attentively to complaints—hear what the specialist says. Many complaints do not state the real problem. Thus, an engineer may complain about the noise in his work area when the real trouble is that his assignment is too difficult for him. He says he'd like to be transferred to a quieter area—what he may really mean is that

he'd like to be assigned to a different and (hopefully) easier task. Or the man who complains about the inadequacy of his salary may really be venting the anger he feels toward his spendthrift wife.

Observe people carefully. Learn to see behind their complaints so you can get at their true reasons for complaining. While this may seem to require a large effort, the results are often rewarding. By hearing what a man says and helping him to overcome his problems, you will convert him to a loyal associate and a happier person. So listen and hear. And remember that many people are unable or unwilling to state exactly what is bothering them. So they turn to kindred complaints, subconsciously hoping that you will discover their true discontent. For as Oscar Wilde wrote, "Discontent is the first step in the progress of a man or a nation."

6. Be kind—most of the time. Kindness is remembered for years. And it costs you nothing. "Kindness gives birth to kindness," Sophocles said. Most engineers and scientists—even the most rugged "management haters"—seek kindness from their supervisors. Why? Because kindness establishes a rapport with the supervisor—makes a man feel that he rates. We all seek this assurance of belonging, of being needed, of having a value as a human being.

Be careful not to allow the professional welcher to make a mockery of your kindness. You can spot this kind of individual easily—he tries to monopolize your time; he takes advantage of your sympathy. You soon find you're spending much of your time and energy on his problems. As soon as you detect a situation like this, exert your authority, and tell the man that you won't be made a sop for his complaints.

Know when to use specialists. You'll work more effectively with specialists of all kinds, within or outside your organization, if you know *when* to call on them. Nothing is so annoying to the well-trained specialist as to be called for help on a task that is far below his capabilities. Here are ten circumstances under which you might seek the aid of specialists or consultants. They are based on recommendations of *Administrative Management* magazine [2] and aim at improving operational efficiency and cutting costs in almost every business operation. Consider using consultants when:

> (1) You don't know what to do yourself. (2) You know what to do but don't know how to do it in your firm. (3) You know what

to do but will have to hire some people and buy more equipment for a short duration job. (4) The job requires special skills which will mean "staffing up." (5) Supplementary manpower would be required which could not readily be absorbed by your firm. (6) You know that an outsider can do the job faster, better, more economically. (7) When the use of an outside firm, particularly for services, can become a guaranteed income for you (as with a lease department, concession, etc.). (8) Department heads recommend using consultants. (9) You wish to test expansion or diversification potentialities. (10) When a decision has such far-reaching effects that it's just good sense to seek another trained opinion.

Obtain maximum help from specialists. The help of a specialist is costly. To derive maximum benefit from a specialist's time, you should strive to get as much assistance as possible from him. Here are five pointers for doing this.

(1) Arrange *in advance* the appointments, equipment, and facilities the specialist will need to accomplish his task. Planning for his visit or assignment will save valuable time. (2) Alert, in advance, all personnel who will deal with specialists. Have these people prepare the documents, data, plans, equipment, and other items that will be needed. (3) Begin working as soon as the specialist arrives. Arrange your own schedule so you can spend as much time as necessary with the specialist. Don't cut him adrift and allow him to wander about, looking for something to do. (4) Check on his progress at regular intervals, but don't watch over him like an overseer. A competent specialist desires and requires freedom to solve the problems presented to him. If you hired him, you should have confidence in his ability. Utilize his time to your advantage. (5) Listen to his recommendations, and do not dispute them until after you've studied them. One of the quickest ways to antagonize a specialist is to disagree with him before you've given his ideas any study.

Specialists Will Always Be with Us

The growing complexity of science and technology means that all of us will have to rely more heavily on specialists to solve the problems that arise in advanced areas of our chosen fields. So getting along with specialists will become a greater necessity in the future.

You can assure your continued achievement by becoming more adroit in your dealings with all specialists. For if you expect to accomplish more, you'll have to do it through the use of the know-how of others. As an engineer or scientist you took pride in what you knew. As a leader of engineers or scientists you'll take pride in knowing how to make maximum use of what others know.

REFERENCES

[1] LeBaron R. Foster, "The Businessman—Through the Eyes of the Intellectual," *Encore*, Spring, 1962.

[2] "When Should You Call in a Consultant?" *Administrative Management*, February, 1962.

State in which you work. Based on the general population one survey [4] shows that the three states having the highest patent productivity are Delaware, New Jersey, and Connecticut. The three lowest are South Carolina, Arkansas, and Mississippi. In between the top and bottom three, New York, Massachusetts, California, Maryland, Colorado, and Iowa rate in the top half. So if you work in one of these states, score your patent chances *excellent.*

Theories, facts, or devices. Check the category that interests you most. You can patent [2] an atomic reactor, spacecraft, diaper pin, locknut, or an electric circuit. But you cannot patent a theory or fact—once revealed anyone can cite your theory or fact and you have little or no legal recourse. The law determines what can be patented—"any new and useful process, machine, manufacture, or composition of matter, or any new or useful improvements thereof. . . ." So if you checked devices, score your patent chances as *good.* Note, however, that you can patent inventions embodying theories and facts.

Are you a "lone" inventor? Many people cite the skyrocketing costs of research and development as a wall in the way of the lone inventor. This is untrue—since 1936 the solitary inventor has managed to earn nearly 40 per cent of all the patents issued.[5] This percentage has remained relatively stable since 1936. So if you work alone on your inventive projects, score your patent chances as *fair to good.*

Male or female? The chances that you're male are very high—if you're reading this book. The same goes for patents. One sample [4] of 5,900 patentees showed only 1.45 per cent were women. So if you're a male, score your patent chances as *excellent.*

How does your score add up? If you scored good or better on half or more of these questions, you have an excellent chance of obtaining a patent should you decide to develop a patentable item. And don't scoff at the value of a patent. More than half the inventions patented in the United States are used some time before the patents expire.[5] Between 55 and 65 per cent of all assigned patents have been in production to some extent at some time.

Do You Know What a Patent Is?

Briefly, if you don't know, a patent joins the United States government and the inventor in an agreement. This agreement stipulates that only the inventor, or someone to whom the inventor grants permission, can sell, build, or use the invention for a period of seventeen years from the date of the granting of the patent. After this period of seventeen years has elapsed, anyone can use the invention. Patent extensions are granted *only* by Congressional action. Since our government represents the people, your patent is actually an agreement with the public granting you the sole right to your invention.

Who Can Apply for a Patent?

You, as the original inventor, your personal representative (patent attorney), or inventors jointly, if there is more than one, can apply for a patent.[5] Two or more persons are joint inventors if they developed the concept while working together. This work can be in the form of research, exploration, or any other method used to develop a concept that is patentable. You can assign or sell your invention to someone else who then becomes the *owner* of the invention. An owner, however, cannot apply for a patent on an invention—the original inventor must do this. The owner can, however, receive the patent as an assignee of the inventor.

How Do You Apply for a Patent?

For best results use the services of a qualified patent attorney when applying for a patent. The investment for his services will be well repaid. You can obtain complete information about the rules and forms for patent applications from the Commissioner of Patents, Washington 25, D.C. by requesting a copy of *Rules of Practice of the U.S. Patent Office.*

Is a Patent Worth the Effort?

It certainly is. Here are three examples of how patents have helped engineers and scientists achieve more in their fields of work.[2]

Dr. Edward H. Land, inventor of the Land camera and founder of the Polaroid Corporation, holds more than 200 patents, and the corporation is a leader in optics and photography. To a large extent, Dr. Land said a few years ago, his success in commercializing and developing his inventions was due to the patents obtained on those inventions. He also said that the strength of the patent picture was largely instrumental in securing adequate capital to finance the Polaroid Corporation and to permit the company to develop its photographic interests and safely spend large sums on research and engineering.

Professor Robert Van de Graff's basic patents for his static generators made possible the formation of the High Voltage Engineering Corporation. Other companies whose patents aided their formation and growth include Raytheon (gas rectifier tube patents) and Electronics Corporation of America.

Jacob Rabinow, president of Rabinow Engineering Co., Inc., was, for many years, an individual inventor. Every clock put into an American automobile today is a Rabinow self-regulating clock. His automatic letter-sorter is being installed in U.S. post offices. And his magnetic clutch, which earned him $100,000 cash for the foreign rights alone, is considered a classic invention.

"Without a patent," Rabinow says, "I would be dead. I would not invent because I could not afford to invent. It took me nine years and several thousands of dollars to develop the self-regulating clock. Without the protection afforded by a patent, I could not have done it. No one wants a free invention. A free patent is a useless patent because anyone can copy it."

Rabinow applies the same principle of patent protection to the inventions made by his firm: If there is no possibility of a patent, even the most promising project is scrapped.

What Can You Do with a Patent?

You have several alternatives, once a patent is granted.[6] Thus, you can (1) make, use, or sell the patented item yourself; (2) sell the patent to an individual or organization; (3) assign the patent to an individual or organization for a stipulated period of time; (4) license one or more persons or organizations to sell, use, or make the patented item; or (5) sell fractional interests or territorial rights to your patent.

As you can see, each of these alternatives holds a promise of financial gain. To most inventors the promise of an income from the sale, manufacture, or use of their patented item is a major incentive.

Many engineers and scientists collect tidy annual royalties from a patented machine, process, circuit, or other development. Having such an income can make a major difference in your personal life. Patent royalties, even though moderate, can be invested for future use. Thus, you might use the royalties to build an educational fund for your children or a larger retirement fund for yourself. And if your patent royalties are large you may, of course, be able to invest in and do many of the things you could not afford on your regular income as an engineer or scientist. So no matter how you approach patents, the return can be both profitable and gratifying.

Stimulate Your Inventiveness

You cannot obtain a patent unless you develop something new and useful. As with anything else in life, inventions seldom develop without work. Regardless of what stories you may have heard of inventors stumbling on million-dollar inventions, the productive inventors are those who work at stimulating their minds to produce new and worthwhile developments. Here are seven hints on how to increase your inventive abilities.

1. Improve your creativity. See Chapter 5 for some useful hints. In general, the greater your creative powers, the better your chances for developing an original and patentable invention.

2. Develop the ability to see a situation from several different viewpoints. A characteristic of the successful inventor is this ability to see many different aspects of a need. One viewpoint may provide you with sufficient information to solve the problem by developing a needed invention.

3. Be alert at all times for patentable ideas. Don't stop looking when you leave work. Continue seeking useful ideas at home, in your hobbies, while on social visits. This constant curiosity, coupled with an eagerness to learn, and a desire to ferret out new ideas may lead you to a profitable patent.

4. Make inventing a hobby. It can be fun and will make your life more interesting. You can profit from this hobby, even if your invention is little more than some gadget to be used around your home. Learn to think in terms of both common and uncommon materials—metal, wood, plastic, paper, cardboard, etc. By combining

these materials, or using one in place of another, you may develop a worthwhile device that is patentable.

5. Use your imagination—solve problems by acting out, in your mind, the exact process a machine, device, or other item would follow in performing a given function. Using your imagination this way will save you many hours because you can eliminate building unworkable models. You will also strengthen your powers of visualization. This ability can aid you in all phases of your professional life.

6. Use your engineering or scientific abilities to develop and improve your basic ideas to the point where they are patentable. Keep working on your idea—even if the development takes years. One worthwhile invention that you patent may repay you more than you might earn in your regular position in ten or twenty years.

7. Work with a reputable patent attorney. Don't try to be your own attorney—the fees you save won't buy many packs of cigarettes. But the losses you may suffer can plague you for a lifetime. So find a good patent attorney and follow his advice.

Your Patents Are Important

You never lose when you obtain a patent. And you may gain much. So begin today to examine your daily activities, hobbies, and home life for patentable ideas. As an engineer or scientist your chances of obtaining a patent are above average. Take advantage of these chances—you may be rewarded beyond your fondest dreams. Learn to think of inventions as Ben Franklin did when asked, "What is the use of this new invention?" Franklin replied, "What is the use of a new-born child?"

REFERENCES

[1] Alfons Puishes, "Patents and the Engineer," *Mechanical Engineering*, August, 1961.

[2] Howard Simons, "Inside the Patent Office," *Think*, October, 1961.

[3] Joseph Rossman, "A Study of the Childhood, Education, and Age of 710 Inventors," *Journal of the Patent Office Society*, May, 1935.

[4] C. D. Tuska, *Inventors and Inventions*, McGraw-Hill Book Company, Inc., New York, 1957.

[5] "Lone Inventor Still Rates High in United States," *Science News Letter*, 80:56 Jul. 22, 1961.

[6] Charles S. Grover, "Patents for Inventions," *Mechanical Engineers' Handbook,* McGraw-Hill Book Company, Inc., New York, 1958.

HELPFUL READING

"Facing Up to a Crisis in Patents," *Business Week,* Nov. 25, 1961.

Hastings, G. S., "Should You Use Outside Inventors?" *Product Engineering,* Jan. 13, 1958.

Kornberg, Warren, "Intensified R & D Aims to Save Patent System," *Product Engineering,* June 25, 1962.

McGraw-Hill News Bureaus, "Foreign Engineers Don't Argue Patent Rights," *Product Engineering,* Jan. 9, 1961.

Szekely, George, "When Was That Patent Issued?" *Product Engineering,* Dec. 5, 1960.

Trimble, Floyd, "What Engineers Should Know about Patents and Claims," *Chemical Engineering,* Oct. 3, 1960.

Wikstrom, Hugo, "Is Your Invention Patentable—and Worth It?" *Product Engineering,* July 17, 1961.

29

Learn the Business Side of Technology

> The fundamental principles which govern the handling of postage stamps and of millions of dollars are exactly the same. They are the common law of business, and the whole practice of commerce is founded on them. They are so simple that a fool can't learn them; so hard that a lazy man won't.
>
> P. D. ARMOUR

Some engineers and scientists are extremely critical of businessmen, saying that these men know nothing. Opinion Research Corporation surveyed 622 engineers and scientists and found that 76 per cent felt that management tries to manipulate people for its own purposes; 77 per cent felt that engineers' and scientists' talents are poorly utilized by management. This schism between technical personnel and management or businessmen has led some engineers and scientists to shun business knowledge. "I'm a specialist," says one engineer. "I'll be darned if I'll give up my specialty to go into the know-nothing area of business."

Yet recent developments in all phases of industry prove the fallacy of this thinking. For example, the two top men at Texas Instruments, Patrick E. Haggerty, president, and John E. Jonsson, chairman, are both engineering graduates. When Haggerty graduated from Marquette University's School of Electrical Engineering, he had the highest grades achieved in the school up to that time. So you see, outstanding technical ability can be combined with effective business skills. Many other outstanding engineers and scientists occupy high positions in business. To name one, Donald R. Lueder, a Cornell Master in Civil Engineering, is president of Geotechnics and Resources, Inc. Mr. Lueder is another example of an engineer who has blended his specialized training with business skill to achieve more in his profession.

Are They "Know-Nothings?"

Let's see if businessmen are know-nothings, as some engineers and scientists claim. The average successful businessman must have at least a rudimentary knowledge of accounting, finance, marketing, advertising, promotion, plant location, distribution, budgets, insurance, taxes, and human relations. In addition, his particular business may require a deep specialized knowledge of a certain field. Thus, a man in the export business must know tariffs, customs regulations, import duties, foreign exchange, etc.

So you see, the average businessman is not a know-nothing. In fact, many successful businessmen have a broader knowledge than some engineers and scientists. Why? Because the businessman works in a larger number of fields. The reason why engineers and scientists sometimes think the businessman is shallow is because many business decisions are vague and difficult to define. There isn't a handy equation, table of allowable loads, or nomogram in sight that will give an answer accurate to a certain number of decimal places. So the businessman seems, to some engineers or scientists, to operate by some sixth sense. This sense—business judgment—is one of the important factors in the growth of the United States to its present position in the world. Also, business sense often develops the capital that provides the jobs for engineers and scientists. Keep these facts in mind the next time you feel the urge to criticize businessmen.

You Need Business Knowledge

You would have much difficulty finding a suitable position as an engineer or scientist if business didn't exist. And the reverse of this, a knowledge of business, can help you no matter what position you may hold in engineering or science. For with a knowledge of business you are better able to understand the decisions made by management that affect your duties, income, and future. You will also have a clearer concept of the place of your industry in the national economy; of how changes in productivity will influence your organization; and how your individual contribution can increase the profits of your firm.

There are personal advantages to you when you have a knowledge of business and business procedures. You are more eligible for promotion to a management position. Your horizons are wider because the business knowledge takes you out of the confines of your specialty.

Stuart Chase, writing in *Think*, said:

> The looming problems of our times demand the development of specialists, yet as the need grows so does the need for generalists—those who are able to take the over-all view, able to relate experience and knowledge in one field to problems in another. It is vital that the specialist himself realize the importance of developing a generalist outlook, for his own sake as well as the world's. As one top executive points out: "While a man's technical knowledge may be his best tool during his first five years with our company, this curve tends to flatten out on the value chart and is met by the ascendent curve of the man's skill in human relations and other factors."
>
> Let the message for specialists be: Live at the level of your time! Crawl out of that talent-trap which you refer to as your "field" and look around. You may learn something about the only era you will ever live in, and about the only species you will ever be a member of! [1]

So resolve today to acquire some useful business knowledge and skills. Such a resolve will never harm you—and with some luck it can help you immeasurably. Let's see how you can acquire this knowledge.

Study business procedures. There are four ways to acquire a knowledge of business: (1) self-study, (2) extension courses, (3) correspondence courses, and (4) by working in the business department of your organization. Each method has advantages, depending on your particular circumstances.

1. *Self-study.* This is the easiest way to acquire an overall view of business. But self-study has disadvantages—it is undirected (you don't have an instructor to guide you); since you don't have exam deadlines to meet, your attention may lag; your study hours tend to be erratic, and without the incentives provided by a planned course you may give up your effort.

But for the man who enjoys studying on his own, self-study can be productive and informative. To acquire business knowledge by self-study: (*a*) Obtain a basic book giving a broad coverage of the field. (*b*) Set up a study plan based on the table of contents of the book. (*c*) Study at a regular pace covering a chapter or section at predetermined intervals—say every two weeks or every month. (*d*) Use the review questions and problems in the book, checking your answers against the text or by an alternative text or computation method. (*e*) Where the book does not contain review questions, prepare a summary of each chapter listing the important facts you learned. (*f*) After finishing the book review it quickly. (*g*) Try to apply your new knowledge whenever you can. (*h*) Select one phase of the subject for further study. (*i*) Get a book on this subject, and repeat the same procedure.

Basic books giving a broad coverage of the field of business include Prerau, Sydney, *J. K. Lasser's Business Management Handbook,* 2d ed., McGraw-Hill Book Co., Inc., New York, 1960; McNaughton, W. L., *Introduction to Business Enterprise,* John Wiley & Sons, Inc., New York, 1960; Hastings, P. G., *Introduction to Business,* D. Van Nostrand Company, Inc., Princeton, N.J., 1961. If you wish to do some studying on your own, any of these books is worthy of your attention.

2. *Extension courses.* These are offered by colleges, universities, and business schools. You can easily attend one of these courses in the evening in your local area. Since these courses have an instructor who requires the use of a specific text and requires the student to pass an examination to obtain credit, they are more in keeping with

your previous educational experience. The various incentives such courses offer may make them the best way for you to learn business principles.

Consider studying for a master's degree in business administration in the evening. Perhaps your organization will reimburse you for part or all of the tuition cost. The M.B.A. degree is popular with many engineers and scientists today for several reasons. Having the degree provides a surer path to management positions of all types. The knowledge acquired while studying for the degree broadens a man's personality and interests. Lastly, a degree in business administration does not usually require as much study time as an advanced degree in engineering or science.

3. *Correspondence courses.* These are also offered by universities, business schools, and home-study institutions and have many advantages. Some universities and colleges even offer college credit for courses completed by correspondence. You must, however, spend a certain minimum time in residence at the university or college before a degree, toward which these credits count, can be granted.

With a correspondence course you have the guidance of an instructor, use recommended texts, and have examinations which enable you to check your progress. Some schools offer a variety of other useful services—advice on solving special business problems, guidance in stockmarket investments, bulletins on business conditions, etc. Since you needn't attend regular classes in the usual correspondence school (unless you want degree credits), you can study anywhere at any time. This is an outstanding advantage of a correspondence school if you live in an isolated area or travel extensively.

Do not, like some engineers and scientists, sneer at correspondence instruction. Some of the most famous business and technical leaders of the United States have, at some time during their careers, studied by correspondence. Here is a list of a few, including one of the important titles each held at one point in his career: R. J. Cordiner, chairman of the board, General Electric Company; Walter Chrysler, president, Chrysler Corporation; Charles R. Hook, chairman, Armco Steel Corporation; Eddie Rickenbacker, chairman, Eastern Air Lines; Charles E. Wilson, president, General Electric Company. For full information on institutions, courses, faculty, and other facilities of various correspondence schools, write the National Home Study Council, 1420 New York Ave., N.W., Washington 5, D.C. See also Chapter 25 in this book.

4. *Working in the business department of your organization.* This is another way to acquire knowledge. But to derive most from this experience, you should study books or courses covering the area in which you are working. Activities in which you can acquire knowledge and experience include finance, accounting, marketing, and business law.

Don't overlook the opportunity to work in the business department even for only a few minutes or hours at widely spaced intervals. Even these short periods will help you acquire useful experience and knowledge. Whenever you work with personnel from the business department, feel free to ask questions. Most people in accounting and similar departments will be delighted to explain their work to you. Besides learning the details of a specific procedure, you may also discover some of the pet methods favored by certain executives. Thus, an accountant may say, "J.B., the executive v.p., wants me to throw in a 2.3 per cent extra supervision charge when figuring the overhead on this new transistorized computer." Knowing facts like these will help you to better understand pricing, write-offs, profit margins, and a variety of other topics.

Read business journals. You can acquire a large amount of useful information about business conditions and practices by reading the magazines and newspapers specializing in this topic. *Business Week, Forbes, Dun's Review, Fortune,* and *Barron's* are just a few of the magazines you will find helpful. Subscribe to one or more of these magazines for a year, and read it regularly. You will learn far more than you ever thought possible.

Newspapers worthy of your reading time include the *Wall Street Journal, Journal of Commerce,* and *The New York Times.* Regular reading of the business and financial pages of these newspapers will keep you well informed about new developments in all the major business areas. These papers are available no matter where you may work in the United States. Buy a copy regularly, or have your name placed on the distribution list in your organization.

Perhaps business is not to your liking—for this reason you chose a career in engineering or science. While actual participation in business may not interest you, a knowledge of business procedures and conditions is a necessary part of your education as an intelligent and informed adult. So recognize this fact today—break out of the confining limits of your specialty and learn what's going on in all phases of business. Read business publications regularly—both the editorial pages and the ads. You'll become a better engineer or scientist, more ably qualified to serve your profession with distinction.

Don't overlook the chance to profit personally from your business reading. For example, regular reading of business journals and newspapers will lead to a better understanding of the stock market. This knowledge, coupled with the background you have in your specialty, may permit you to make a profitable investment in stocks, bonds, or other securities. Some engineers and scientists have been able to accumulate a substantial reserve of funds by coupling technical and business knowledge. But before you begin to spend the money you dream of acquiring this way, get into the habit of regularly reading good business literature.

Your Own Business

American engineers and scientists, like many of their fellow citizens, often dream of a business of their own. And since the end of World War II probably more graduate engineers and scientists have gone into business for themselves than during any comparable period in our history. All current studies and perdictions of future business activity indicate that *more* engineers and scientists will set up their own businesses in the next few years. If you've ever dreamed of a business of your own in some area of technology, your chances of ever forming the business and making a success of it, are much greater if you have a good knowledge of business procedures and practices. Just as you'd laugh at a meat butcher who tries to be an electronics engineer without any specialized training, so too would a businessman smile at the thought of the engineer or scientist who goes into the marketplace without training.

Business procedures seem heartless and ruthless to some engineers and scientists first getting into the marketplace. As a technical specialist you are often protected from the rough and tumble world of pricing, profit margins, legal squabbles, contract cancellations, etc. Many engineers and scientists frown on the term "business decision," feeling that it represents a compromise of intellectual integrity. This may or may not be true—depending on the circumstances. Regardless of your feelings about such matters you have little chance of succeeding in your own business unless you adopt a businesslike attitude. You can help overcome your distastes for the ways of the marketplace by learning more about the ethics of business. "Knowledge is power," as the sage said. This proverb applies as aptly to business as it does to engineering and science.

Let's for a moment look at some engineers and scientists, and the successful companies they formed. Here are five,[2] picked at random from a list of several hundred:

> (1) James E. McClain, an electrical engineering graduate of the University of Texas, founded Electrical Service, a company that overhauls, repairs, and up-rates used transformers. His firm also builds transformers for special duties. (2) Dr. John L. Barnes, a Princeton University Ph.D., founded Systems Laboratories Corporation, specializing in lunar flight, guided missiles, flight control, and similar tasks. (3) Charles Bartley, a graduate of the University of Maine with a B.S. in engineering physics, was founder of the Rocket Division of the Grand Central Aircraft Company. The division, with Bartley still in charge, was later made a separate company, Grand Central Rocket Company. (4) Arthur O. Black, an M.I.T. electrical engineering graduate, founded Magnetics, Inc., to manufacture magnetic cores, amplifiers, reactors, and a number of similar products. (5) Wilbert Chope, an Ohio State electrical engineer, founded Industrial Nucleonics Corporation, to produce radiation thickness gauges, quality-control systems, and related equipment.

All these men, and hundreds of others, are successful in their business-technical careers. The editors of *Fortune* magazine explain the lure of forming an engineering or scientific business in the following way:

> Most scientists are far happier doing research in their secluded laboratories than they are as manufacturers competing for sales and government contracts. The technical expert turned entrepreneur is therefore a rather rare type. But they are also among the most interesting because, when they apply pure logic to business ventures, they frequently get unexpectedly good results. The best of them combine technical brilliance with salesmanship and the managerial ability that many a big corporation would be happy to have. But very seldom does the scientist-turned-businessman leave his

small corporation to return to a laboratory where he can apply his talents to pure research. It is as though, having mingled with the dollar-chasers, he has found money-making a fatal, if unscientific attraction.[2]

You *can* go into business for yourself, if you have the desire and interest. The best way to build the desire and interest is to learn the basic principles of business. You can learn these principles by following the hints given earlier in this chapter. Start today and see how fast a knowledge of business coupled with technical skills can increase *your* professional achievements. Remember: you'll never know until you try.

Business Knowledge Pays Off

Let's conclude this chapter with another look at what business knowledge can do for you. Business skills, when well-developed, can (*a*) improve your thinking ability; (*b*) make you a better speaker, more at ease in any gathering; (*c*) improve your writing skills to the degree where you can write technical or business reports, memos, and letters with ease; (*d*) sharpen your human relations techniques, making you a warm individual, welcome wherever you go; (*e*) increase your income to a level far beyond that possible in many a technical specialty; (*f*) give you a chance to develop and successfully operate a business of your own; (*g*) broaden your analytical ability; and (*h*) expand your personality, giving you wider interests and better understanding of your fellow men. What more can any engineer or scientist ask for? Remember that know-nothing businessman you've heard some technical people lambast? He's more knowledgeable than he appears to be, isn't he?

REFERENCES

[1] Stuart Chase, "Are You a Specialist or a Generalist?" *Think*, August, 1960.
[2] Editors of *Fortune, Adventures in Small Business,* McGraw-Hill Book Company, Inc., New York, 1957.

HELPFUL READING

Allen, L. A., *Management and Organization,* McGraw-Hill Book Company, Inc., New York, 1958.

Anderson, R. C., *Management Practices,* McGraw-Hill Book Company, Inc., New York, 1960.

Cordiner, R. J., *New Frontiers for Professional Managers,* McGraw-Hill Book Company, Inc., New York, 1956.

Dale, Ernest, *The Great Organizers,* McGraw-Hill Book Company, Inc., New York, 1960.

Gray, J. S., *Common Sense in Business,* McGraw-Hill Book Company, Inc., New York, 1956.

Grimshaw, Austin, *Problems of the Independent Businessman,* McGraw-Hill Company, Inc., New York, 1955.

Heimer, R. C., *Management for Engineers,* McGraw-Hill Book Company, Inc., New York, 1958.

Houser, T. V., *Big Business and Human Values,* McGraw-Hill Book Company, Inc., New York, 1957.

Keith, L. A., and C. E. Gubellini, *Business Management,* McGraw-Hill Book Company, Inc., New York, 1958.

Koontz, H. D., and C. J. O'Donnell, *Principles of Management,* McGraw-Hill Book Company, Inc., New York, 1959.

Maynard, H. B. (ed.), *Top Management Handbook,* McGraw-Hill Book Company, Inc., New York, 1960.

Murphy, T. P., *A Business of Your Own,* McGraw-Hill Book Company, Inc., New York, 1956.

Prerau, Sydney (ed.), *J. K. Lasser's Executives' Guide to Business Procedures,* McGraw-Hill Book Company, Inc., New York, 1960.

30

Be a Part- or Full-Time Consultant

> The shoemaker makes a good shoe because he makes nothing else.
>
> EMERSON

You have a wonderful adventure ahead of you once you decide to become a part-time or full-time consultant in any engineering or scientific field. To consult in any field of engineering, you will need your professional engineer's license. Consultants in scientific fields—physics, chemistry, metallurgy, and a host of others—do not normally need a state license, other than a business license in some localities. If you're an engineer, you should have your professional license anyway, regardless of whether you intend to consult or not. See Chapter 24. As a scientist you can operate freely, but having a professional engineer's license may permit you to accept some clients you couldn't otherwise. So why not check into the possibility of obtaining a license in your state?

Why Consult?

There are many reasons. A few of the more important ones include increased professional knowledge, larger income, more varied professional contacts, a chance to contribute more to your field, increased personal satisfaction from your work, greater independence—as a full-time consultant you avoid the nine-to-five regularity. Most of these advantages accrue to you even as a part-time consultant. The added income from your part-time work leads to a better life for yourself and your family. You meet new people, expand the scope of your knowledge and friendships. You will even improve your human relations and leadership skills. All this extra experience will make you a better performer in your present position and can open horizons you never dreamed of. So read this chapter carefully—it may open a whole new way of life to you.

Who Can Consult?

Almost any well-trained engineer or scientist who has a deep knowledge of one phase of his field can become a consultant. Of course, some areas of technology and science are easier to enter as a consultant than others. Why? Because there is a greater need for consulting experience and help in certain newer specialties than in other, longer established specialties. Thus, you could probably find more outlets for your services in space and missile specialties today than you could in textile technology.

Are there any specialties that are too narrow for a consultant? There are very few. In fact, the narrower your specialty the greater your knowledge. And since consultants are generally hired because they possess superior knowledge of a given subject, your chances of finding clients increase with the degree of your knowledge. Charles Lynch, while assistant editor of *Product Engineering* magazine conducted an interesting survey [1] of consultants. Here are the helpful facts he uncovered.

What It Takes

"There's a race of men that don't fit in," wrote Robert W. Service, the frontier poet and backroom ballader. And among engineers and scientists, independent consultants perhaps belong to this breed. It's more than being above average in talent. Their motives are different ... their attitude is different ... their goals are different. In these times when most other engineers and scientists seek the security and stability of a large corporation, these rebels dodge group creativity, team-oriented business, and a slot in the organization chart. They prefer a basement-size engineering department, independence and the risks of private enterprise. Specialized consultants aren't thirsty for expense accounts or fringe benefits, and sermons on teamwork leave them cold. But, they say, the advantages of going it alone far outweigh the disadvantages—if you've got what it takes.

"I'm a kind of maverick," says one as he tips his chair away from his workbench and clasps his hands behind his head. "I don't work a 9-to-5 day, and of course don't get that regular pay-check at the end of the month. My fringe benefit is the satisfaction of doing the work I want to do, the way I want to do it. It's flattering to think there are companies who will pay me so I can live this way. Some people wouldn't like it. But to me it's exciting—I never know who will call me next or what his problem will be."

The large consulting firm, of course, is not a novelty in engineering and science. Large consulting firms have been around a long time. These are the corporations that supply engineering talent to build bridges and vehicular tunnels, install power plants and hydroelectric projects. These are consultants in the classical sense—as familiar to the public as they are to the engineering profession.

The Engineer's Engineer

But in this age of specialties, a new kind of consulting engineer is beginning to emerge. He is the independent who, through publicity in engineering journals and society meetings, has acquired a reputation as an expert in a narrow field of engineering—and has capitalized on it. He may rent a room in an office building or he may operate out of his own home; he may earn much money . . . or he may starve.

Independent consultants can be grouped broadly into two classes. There are those to whom consulting is only a part-time job—perhaps a natural appendage to their work as university professor. They don't rely on consulting for a livelihood—it's simply a welcome and challenging source of extra income. This type of consultant isn't subject to the fluctuating whim of supply and demand—their salaried main job is a hedge against variations in the freelance market.

The other type of consultant is an all-out entrepreneur. His sole source of income is the consulting service he can sell. In his firm he is the president, vice-president, sales manager, engineering staff, bookkeeper, and shipping clerk.

Sheer Engineering Competence . . . Plus

Versatility, and a creative approach to unusual problems, are the attributes that distinguish the independent consultant from his fellow engineers and scientists. But what other qualities are needed? Engineering or scientific competence, of course, is basic. To the independent consultant this means experience, years of it, and a reputation in his established field. For this reason younger engineers and scientists have a harder time of it than the older, better established man. But a young man in one of the new, fast-growing fields can command respect too.

Salesmanship appears to be the second most desirable quality. "The ability to sell a customer and keep him sold makes the difference in whether the consultant will eat tomorrow," a gear expert explains. Then he adds, "It is also the quality most likely to be lacking in an engineer."

These traits—versatility, engineering or scientific ability in a chosen specialty, and sales ability that flows from an infectious enthusiasm—are perhaps the only qualities found in common among these individualists. Their specialities, work habits, even the ways they earn money or contact clients, vary from man to man. As might be expected from a group unfettered by the red tape and controls of a larger organization, they find the pattern that suits them best and bet their brains and bank account that the services they offer will be marketable.

Easy Does It

Putting yourself in business isn't nearly as simple as it looks. To hang out your shingle and wait for someone to knock on your door, said one man candidly, "is probably the most certain, and surely the dullest path to bankruptcy you could choose." Advertising is not much good either (aside from the matter of dubious ethics). Most independents don't even bother to list themselves in the yellow pages of the phone book. The field of independent consulting is so small, so new, and so fluctuating that most of the men aren't even listed with their engineering society.

Usually an engineer or scientist gets into the consulting business slowly and cautiously—almost accidentally. The first job may come through a friend or business acquaintance and is done on a part-time basis. Then, somewhere during the middle of the second part-time job, the consultant starts feeling that on the strength of his reputation and the broad application of his specialty, consulting work will support him full-time. (But watch out, warn those who have tried it; the first two years are the hardest and it's best to have some source of independent income.)

After he becomes a little better established, he will probably want to keep several jobs going simultaneously. Contacting new clients may develop into a sizable part of his job. Friends, previous employers, vendors on a previous job, engineering and scientific societies are all good sources of new work. Ringing doorbells of companies having products related to his specialty is another common technique. Fortified with a scrapbook of published papers and lists of his patents, he hopes to be placed on the list of qualified consultants to be considered when a problem in his specialty arises. Some independents send reprints of their by-lined articles to a carefully chosen list of potential clients. One man has an interesting gambit. He scans employment ads for companies trying to hire full-time employees for a job he could do part-time—then sells the company on the idea.

Keeping That Wolf from the Door

Most well-established one-man consulting firms have at least one or two steady clients who pay them on a retainer basis. A research physicist, for example, is on retainer to two large corporations which he visits perhaps once or twice a year at most. He spends the rest of his time at home, in his laboratory—or at his desk doing undirected research which may or may not lead to something profitable for his two clients. Does he do work for other companies too? "Well," he says, "my clients don't deny me that opportunity, but I don't invite any more work. I can keep busy just working for these two."

Another man, a metallurgist, works for 14 different noncompeting companies which he visits once or twice a month. His office is in the basement of his home located in an attractive suburb. He's been an independent consultant 12 years. In his spare time he teaches; writes books and technical papers. He works hard; lives comfortably.

Why does a company hire an independent consultant? First, and most common reason: A small company requires specialized knowledge to supplement its engineering effort but can't afford to hire a specialist full-time. Even large companies, with many such specialists already on the staff, often need an outsider with recognized competence. The consultant offers availability, a fresh viewpoint, and sheer engineering competence derived from years of experience with many wide-ranging problems in a narrow specialty.

The other common reason: Extra help when the engineering department is overloaded. Most consultants prefer to avoid this type of work—it's too confining, too much of the routine they were trying to avoid when they launched into private practice. But for the employer, they are a good buy in such situations. Independent consultants involve no overhead, no fringe benefits, and are not a permanent addition to the payroll to be carried when the need is gone. On an hourly-wage basis they are extremely expensive, but this is not a fair comparison—the full-time staff engineer costs more than twice his base pay in overhead.

Baggy Pants No Liability

When called in on a job, consultants may be put to work in several ways. Those who make regular visits on a retainer usually spend a day talking with the men in charge of various projects. The job of these consultants is to keep efforts on the right track, point out new directions, and assure management that money is not being drained away on projects with little merit. Sometimes the consultant is given an office for the day and is visited by people bringing technical problems. "It's a rewarding experience," says a servo-system expert. "Maybe you were up nearly all night on a previous job. You have a two-day growth of beard, your pants are baggy, shirt open at the neck, but they don't care—you're an expert. They literally line up outside the door. Vice-presidents run to buy your coffee. They pick your brains, you tell them what you know and they are apparently grateful. At least they continue to pay you to come back."

Another common arrangement is to be called in on a particular problem. It may be to design a small gearbox for a particularly ticklish application. In such cases the consultant gets an exploratory contract—this gives him a chance to make some preliminary calculations and estimate what the bill will be. "Performance is what the company wants," consultants stress. "They want that gearbox and they want it within a specified period of time. They don't care where you get the information—books, manufacturer's data, other consultants—it doesn't matter. Performance is what they want and what they pay for."

Because many clients like to get a production prototype on completion of the job, some consultants find it convenient to maintain a small machine shop and testing laboratory. And those who have invented a specialized device with a limited market may also have a small production shop. In some cases, the consulting service functions mostly as a sales promoter. If the application calls for it, the consultant may recommend his patented device and offer to custom-tailor it to the job.

Writing articles for trade publications and technical journals is not a very fat source of income. But the consultants do it—mostly to maintain status in the field. Or, in some cases, the consultant will write an article relating to some specialized work he has done for a company. The article is then published under company auspices, for company publicity; the consultant collects the standard author's fee plus a handsome bonus.

Growing Pains

The consultants are first to admit they are not working in utopia. It requires an unusual kind of personality to get into the business and make it pay, they say.

Here are fourteen of the items established consultants list when questioned about factors they underestimated or mistakes they made early in their practice:

1. It takes time to get established. There will be many difficult moments the first few years. It's best to have saved some money, or have some source of independent income to carry you through the slack months.

2. Don't forget that expenses are your own. If you take a trip to a

society meeting, it comes out of your own pocket. And there will be no group insurance plans or provision for sick leave either.

3. A one-man firm may have difficulty getting catalogs, company literature, trade magazines. A time-clock man can usually get such things through regular channels.

4. Don't spend so much time on the project that you lose touch with what the rest of industry is doing. The facts a consultant gleans from a wide variety of sources are his stock in trade.

5. There are times when you won't get the personal satisfaction of finishing a job you've started. If your recommendations are modified or misinterpreted after the situation has passed out of your control, you are blamed anyway.

6. Don't undercharge for your services. Operating costs are greater than they appear at first glance and your work-load has cyclical fluctuations. You are a specialist and you should charge accordingly. Fees for consultants vary and are, of course, tied to the general economic picture. A daily charge of $150–$300 is common.

7. Don't overestimate the potential market. One of the first things a company cuts when it begins to get economy-minded is service of this type. During periods of economic stress, the company may want to continue the relationship on a minimum basis but this is not very satisfactory. The consultant may not mind starting on a minimum basis—this is customary. But he anticipates the company will want more of his time when it gets deeper into projects he has suggested. If he is a top-notch man and has done a good selling job, he needn't worry.

8. "Ventilate" your working schedule—allow a day or two between consulting jobs so you can catch up on new catalogs, the latest issues of key magazines, and other developments in your field. This will help you recharge your knowledge reservoir so you're better equipped to handle your next consulting job.

9. Don't become any type of consultant—part-time or full-time—until after you have enough knowledge of a given field to call yourself a specialist.

10. Publicize your entrance into the consulting field by issuing news releases to the magazines and professional journals in your field. Figure 30.1 shows a typical example of a published news release announcing a consultant setting up shop.

11. Don't be chary on other forms of publicity—business cards,

brochures, article and paper reprints, résumés, etc. Design, or have designed, a neat letterhead. Use the same design for billheads, cards, catalogs, and all other literature you distribute. A good design will help create a favorable image for your business.

12. Develop greater interest and activity in your professional society. You will obtain more job leads from the society than you ever imagined. See Chapter 26 for helpful hints.

13. Remember—don't discourage easily. Starting a new business is difficult for even the most talented people. Engineers and scientists are no exception. So keep plugging until you have enough clients. That big client may be closer than you think. But be prepared: Learn the other aspects of your business—simple accounting, bookkeeping, tax laws, record-keeping, etc. Go all out to succeed and your chances are splendid. And if you're that rare breed of engineer or scientist we mentioned earlier, you'll love every minute of your consulting work.

14. Be on the lookout for useful fill-in tasks like part-time teaching, writing, lecturing, conducting seminars, serving as an expert witness for lawyers. These activities can publicize your name, broaden your knowledge, and stabilize your income.

> **Rodney D. Chipp Heads Consultant Group**
>
> UNTIL recently an engineering executive at International Telephone & Telegraph Corp., Rodney D. Chipp has left to head the new firm of Rodney D. Chipp & Associates, Consulting Engineers, with headquarters in Bloomfield, N. J.
>
> Besides pioneering in television and radar, Chipp has three decades of experience in design and operation and management of electronic and communications systems.

FIG. 30.1. Announcement of formation of a consulting engineering firm as published in a magazine. (*Electronics*)

REFERENCE

[1] C. J. Lynch, "One Man Engineering Firm," *Product Engineering*, Feb. 23, 1959.

HELPFUL READING

American Society of Mechanical Engineers, *Handbook of Consulting Practice for Mechanical Engineers*, American Society of Mechanical Engineers, New York, 1961.

Bullinger, C. E., *Engineering Economy*, 3d ed., McGraw-Hill Book Company, Inc., New York, 1958.

Canfield, D. T., and J. H. Bowman, *Business, Legal, and Ethical Phases of Engineering*, 2d ed., McGraw-Hill Book Company, Inc., New York, 1954.

Constance, John, *How to Become a Professional Engineer*, McGraw-Hill Book Company, Inc., New York, 1958.

Dunham, C. W., and R. D. Young, *Contracts, Specifications and Law for Engineers*, McGraw-Hill Book Company, Inc., New York, 1958.

Janney, J. E., "What Makes a Business Builder?" *Dun's Review and Modern Industry*, January, 1959.

Marston, A., Robley Winfrey, and J. C. Hempstead, *Engineering Valuation and Depreciation*, McGraw-Hill Book Company, Inc., New York, 1953.

Mead, D. W., H. W. Mead, and J. R. Akerman, *Contracts, Specifications, and Engineering Relations*, 3d ed., McGraw-Hill Book Company, Inc., New York, 1956.

Nord, Melvin, *Legal Problems in Engineering*, John Wiley & Sons, Inc., New York, 1956.

Sadler, W. C., *The Specifications and Law on Engineering Works*, John Wiley & Sons, Inc., New York, 1948.

Stanley, C. M., *The Consulting Engineer*, John Wiley & Sons, Inc., New York, 1961.

Stewart, H. F., "What It Takes to Be Your Own Boss," *Think*, April, 1961.

Tucker, J. I., *Contracts in Engineering*, McGraw-Hill Book Company, Inc., New York, 1947.

31

Make Professional Success a Habit

> For the most part I do the thing which my own nature drives me to do. It is embarrassing to earn so much respect and love for it.
>
> ALBERT EINSTEIN

If you've read this book carefully up to this paragraph, you know more fully than ever before that success and achievement in engineering and science can be yours. But you must work to achieve worthwhile goals. Often this work involves skills outside the normal ken of engineering and science. Once you recognize this and pledge yourself to acquiring the new and needed skills, greater achievements are within your reach.

How can you make professional achievement a habit? It's easy—if you apply the hints given in this book. To permit you to quickly review some of these hints, this chapter is designed as a rapid refresher of much of what you've already learned in earlier chapters. Also, it is a place of resolve where you can reinforce your knowledge of the earlier chapters and reaffirm your intention to put that knowledge to work. Here are twenty habits for professional achievement that will pay off for you in every field of engineering and science.

1. *Know your goals.* You'll achieve little, unless you know what you're trying to accomplish and how you'll do it. So choose your goals. Then plan each step you'll take to reach a specific goal. Write your plan out, numbering each step. Check off each step as you complete it. Watching the list become shorter will spur you to greater achievements.

Careful planning is the key to accomplishing more. Having plans directed at achieving a specific goal helps you channel your efforts so you achieve more in less time.

Once you achieve your first goal, set up another. Keep a goal in view at all times. This technique will lead you on, step-by-step, to greater achievements in every part of your professional career. As Publilius Syrus wrote, "Do not turn back when you are just at the goal." And as Justice Oliver Wendell Holmes said on his ninetieth birthday, "The riders in a race do not stop short when they reach the goal. There is a little finishing canter before coming to a standstill. There is time to hear the kind voice of friends and to say to one's self: 'The work is done.' But just as one says that, the answer comes: 'The race is over, but the work is never done while the power to work remains.' "

2. *Be the master of your time.* Don't waste minutes—or the hours will slip by too. Control the time you spend on professional work—don't allow people to crowd your day with useless talk, unimportant details, sub-professional tasks. Take time to be friendly, considerate, and helpful. But also take time to be productive, efficient, effective. Plan to use your odd moments—travel time, reception-room waiting time, spare time. Once you've made time plans, put these plans to work. Be master of your time, and save time by making quick, accurate decisions. Don't diddle—produce. You have much to gain, nothing to lose.

Cut idle thoughts to the minimum. Set your mind to working on a problem—professional or personal. Remember—as an engineer or scientist you always have something constructive you can think about or work on. C. P. Snow said, "Scientists regard it as a major intellectual virtue, to know what not to think about."

3. *Organize your workday for efficiency.* Look forward to every day as a new chance to achieve your goals. Start early and keep working. Take an occasional breather to refresh yourself. Train the people you supervise to observe your working routine. Don't allow them to impose on your time. Set certain hours aside for interviews, reports, meetings, and discussions. Reserve the other hours in the day for dictation, telephoning, and routine tasks. Get to know when you think best. Do your creative thinking and planning at that time.

Arrange your desk, files, correspondence, and routine paper work for maximum efficiency. Learn to use your wastebasket to unclutter your desk, your mind, and your life. Don't be afraid to try new ways to organize your workday. Just because you've been doing something one way for five years does not mean this is the best way. Look for new ways to improve the quantity and quality of your daily output. When you find an improved method, put it into use. If it saves your time and energy without reducing the quality of your work, adopt the method as a permanent procedure.

Learn to make full use of all the modern aids to business—your telephone, dictating machine, computers, teletype, telegrams, air mail, special delivery mail, etc. By adroit use of these and other aids you can often save time, long trips, and wasted effort. Plan your workday so you will have time to use these aids when you need them.

Careful planning and organization of your day can help you achieve more in less time. It will also free your evenings and weekends so you can relax properly. Spending your free time with your family and hobbies will refresh you, making you better able to achieve more in your profession during your working hours.

Remember what Auren Uris, well-known business writer of Research Institute of America said,[1] "... the problem of time-saving is one not so much of method as of *motivation*. Almost any systematized approach to better time utilization would improve the effectiveness of the average manager."

4. Think—and act—like an executive. "Ninety per cent of an executive's time is spent in the selection, development and supervision of others. About ten per cent of his time is spent on 'factual' effort. These percentages are probably just exactly reversed for the engineer. This radical change in daily endeavor must be anticipated and understood." [2]

Your promotion potential,[3] according to Detroit University's Dr. T. R. O'Donovan, depends on: (1) your ability to increase your capacity to see your job and department in the light of your company's total operations, (2) your ability to get along with others, (3) your ability to develop confidence in your present position, (4) your ability to keep your skills up-to-date through education and training.

Begin now to think and act like an executive. Try to understand the importance of achieving results through the direction of other people. Recognize that the businessman and manager are important to you and your career and to every organization operated by men. Rid yourself of the petty jealousies and unreasoning envy so many engineers and scientists burden themselves with. Study the meaning of profits; recognize that profits create jobs—your job. Keep expanding your horizons beyond the limits of a narrow technical outlook. Stop complaining about how low your salary is compared to truck drivers, carpenters, etc. Get off your comfortable swivel chair, and do something that will rate you a better income than the truck driver.

Power magazine's popular associate editor, Steve Elonka, once summarized ten traits that hold the key to success. These traits are: (1) drive, (2) broad interests, (3) ability to organize, (4) ability to rise to challenges, (5) self-reliance and basic security, (6) ability to be realistic, (7) intelligent use of time, (8) stamina, (9) originality—ideas, and (10) knowledge. Note—*drive* comes first. So put purpose in your life by imbuing it with drive for worthwhile goals and important achievements. As Karl von Clausewitz wrote, "The best strategy is always to be very strong, first generally, then at the decisive point."

5. *Develop your creative abilities.* "Engineering work has assumed a greater position in industry and, therefore, there is greater opportunity to do significant, creative work and obtain greater compensation."[4] This statement, from a graduate engineer, shows the importance most engineers place today on creative ability in every phase of their work.

Creative ability pays off in all phases of your life—professional and personal. So learn the basic facts about creativity, and then apply them to your life. Even if you do no more than develop a more convenient way to get from your home to work, your creative efforts will be worthwhile. But with engineering and science advancing at such a rapid rate, even your most tentative creative efforts will produce some beneficial results.

A creative outlook stirs every engineer and scientist to a greater awareness of himself, his profession, and his special interests. This awareness is almost certain to lead to some useful professional developments. Developing your creative abilities, and we all have some, prevents you from being like the man Alfred Polgar had in mind when he wrote, "When man at the end of the road casts up his accounts, he finds that, at best, he has used only half his life, for good or bad purposes. The other half was lost inadvertently, like money dropped through a hole in the pocket."

Hans Zinsser outlined one job for creative thinking that applies to every engineer and scientist. He said, "Our task as we grow older in a rapidly advancing science, is to retain the capacity of joy in discoveries which correct older ideas, and to learn from our pupils as we teach them. That is the only sound prophylaxis against the dodo-disease of middle age."

6. Don't shun problems—solve them. There is a tendency, among some engineers and scientists, to shun problems once school is out. Yet your greatest asset in life is the ability to face and creatively solve difficult problems. Why squander such a valuable asset just to avoid thinking? For, after all, we are avoiding thinking when we shun problem-solving.

Practice taking on difficult problems in your professional activities. Forget what may happen if you fail to solve a problem. Concentrate, instead, on the benefits that will be derived if you solve the problem. Follow [5] Harold E. Brewer's six steps of problem solving: (1) define the problem, (2) establish objectives, (3) get the facts, (4) weigh and decide, (5) take action, and (6) evaluate the action.

Seek to solve every problem you meet. Don't be discouraged if you're only half successful. Certainly half a solution is better than none. And with half a solution your road to the complete solution may be far easier than you've ever imagined. Remember the words of an outstanding engineering executive. He said, "Show me a man who can consistently solve the problems he faces and I'll build an entire industry around him. For what this world lacks most are engineers and scientists who delight in solving the tough problems of technology and science."

7. Make your memory work for you. Don't be hobbled by a misfiring memory—it can cause you loss of time, money, and energy. Use the simple hints given in Chapter 7. Apply the findings reported by Sterling D. Huggins in *Advanced Management* magazine. He states, "We generally remember 10 to 15 per cent of what we *hear*,

15 to 30 per cent of what we *hear and see,* 30 to 50 per cent of what we *say,* 50 to 75 per cent of what we *do,* and 75 per cent *of what we do under proper supervision and coaching.*"

A good memory can make you stand out in a crowd—most people are so woefully neglectful of their memory abilities that they admire the man who takes time to remember. As Louis Nizer said, "All of us have extraordinarily good memories and also very bad memories. It depends on the degree of concentration we bring to our memories."

Never underrate the importance of a good memory. Being able to recall accurately will serve you in hundreds of circumstances in all phases of your professional career. Thus, you will be able to recall details of technical conferences, the names of important people, sources of key publications in your specialty, contents of letters, reports, and other documents, appointments, etc. Alexander Smith wrote, "A man's real possession is his memory. In nothing else is he rich, in nothing else is he poor."

Lastly, if you think you have a poor memory, you're in for a surprise: Your mind can store 600 memories *per second* for a lifetime of 75 years *and still feel no strain.*[6] The key to a good memory, says the New York Telephone Company's *Telephone Review,* is a simple one—*use it.* Here are some tips for keeping your memory limber: (1) *Concentrate* on what you want to remember. *Be selective* about what you want to remember. (2) *Find a reason* for, and some *interesting* aspect for, remembering. (3) *Repeat what you want to remember.* (4) *Don't write it down.* If you make a note of something in writing there is obviously no reason why you should burden your memory with the same information. (5) *Visualize an association* around a notable feature. (6) *Fit memories into a pattern.* When you have to remember a string of items, fit them into a pattern that can be opened with a key.

8. *Use planning to advance your career.* Learn to plan effectively in all phases of your life. A plan provides you with a road map to the future. Keep your plans flexible. Publilius Syrus warned, many centuries ago, "It is a bad plan that admits of no modification."

Use plans to project your future actions. To use plans for this purpose: (1) Choose a course of action to achieve a specific objective. (2) Plan each step you will take—list these steps in an orderly fashion. (3) Mentally imagine you have taken each step and try to forsee what problems or difficulties you will face and how you will overcome them. (4) Objectively compare the probable outcome with the desired results. (5) If necessary, alter your plans to bring the results closer to the desired objective.

Nothing is so empty as a day without a plan. Begin now to make plans work for you. For as Edward R. Tarrabee said, "No matter how lofty your ambitions may seem, do not ridicule or stifle them. Live them. Sleep them. Dream them. Keep on and on and your ambitions will come true." Good plans lead to achievements in every area of your life.

9. *Build your writing skills.* The engineer or scientist who writes well is a rarity. Because so many of your fellow professionals disdain writing, you can become a standout by simply learning how to write clearly and effectively. Certainly writing is work—but nothing worthwhile in life is accomplished without work.

Epictetus said, "If you wish to be a writer, write." This is probably the best advice anyone could give you on the art of writing. Write to express your ideas clearly and concisely, not to impress your readers with your deep knowledge of the subject. For if you are well-informed about your subject and write clearly, your knowledge will show through. Your readers will praise you for your clear thinking *and* deep knowledge.

Approach each writing task as a challenge to your skills. Try to make each written piece better than the last. Teach yourself to write under all conditions—noise, weather, place, physical feelings. Once you've learned to write in a noisy atmosphere, in bad weather, or when you feel ill, you can call yourself a professional. For this, essentially, is what the professional writer can do. He has taught himself to produce acceptable copy regardless of the external situation. With this ability he combines a facility to express his thoughts clearly and concisely.

Every engineer and scientist has the opportunity to remove the "terrible writer" stigma from the profession. How? By making every effort to see that he avoids the pompous verbosity that has saddled our profession for so long. So resolve now to write better—for your own advantage and the profession's.

10. *Have a hobby—and pursue it.*"No man is really happy or safe without a hobby," said Sir William Osler. You work hard, Mr. Engineer or Mr. Scientist, in making a contribution to your field. You also work hard to provide for your family. But these two—your profession and family—do not give full meaning to your life. There must be a third element—a hobby—that fills your thoughts and plans during your idle moments.

So find a hobby that fills your spare time. Pursue this hobby with gusto—it will delight you, refresh you, make you better able to contribute more to your profession and family. Don't listen to the man who says, "My hobby is my work." He's usually a bore because he has nothing to rejuvenate him mentally and physically. Like a steady diet of any food, all work can dull the senses and make a man a dolt.

11. *Recognize the importance of money.* Many engineers and scientists have a detached attitude toward money. They approach a new job less from the standpoint of earnings than from "what they'll be doing." Certainly, a detached attitude toward money is not a serious criticism to level at a man. Yet if such an attitude limits a man's earnings, both he and his family may be deprived of certain benefits. Also, a failure to recognize the importance of money can limit a man's achievements in business activities because he overlooks profits and opportunities to negotiate profitable transactions. Thus, the engineer or scientist is likely to give all his thought to how a new product works instead of spending some time thinking about how the product will be received by the purchasing public.

Recognize that money has its place in your professional life. Without the capital to build engineering and scientific facilities, employers of technical personnel would have no jobs to offer engineers and scientists. Without jobs few of us could practice our chosen profession.

So cast off your disdain of the manager and his attention to money. True, the educated intellectual, represented by most engineers and scientists, may have more rigorous training than the manager in certain disciplines. But this is no reason to disdain the manager. Don't be like one intellectual quoted in a recent Opinion Research Corporation survey [7] who said, "The intellectual looks down on the businessman as an inferior breed of human. And the businessman looks at the intellectual as a queer kind of person who is soft in the head and best avoided."

Lastly, when recognizing the importance of money, keep George Horace Lorimer's famous remark in mind, "It's good to have money and the things that money can buy; but it's good to check up once in a while and make sure you haven't lost the things that money can't buy."

12. *Get along with everyone—even if it takes patience.* As with money some engineers and scientists have a detached view toward human relations. "Why bother," they ask, "when the design, construction, or some other technical aspect of a project is the really important factor?" The reason for bothering is that people are a fundamental part of every project. You will seldom achieve much as a loner; consistently achieving greatness usually requires that you work *with* and *through* others. But you don't want to do great things, only moderately successful things? You must still work with people—in fact you can hardly avoid doing so in today's economy.

In working with other people, keep these hints [8] in mind: (1) Decide in advance what you will contribute; (2) be careful not to step on toes; (3) give credit where it is due; (4) prepare a schedule; (5) agree upon a coordinator; (6) keep a record of what is said and done; (7) give each person involved a special motive; (8) prove that your project benefits the organization; (9) expect to compromise and plan accordingly; and (10) anticipate objections to your ideas and be willing to accept the suggestions and ideas of others.

If you get nothing more from this book than a greater appreciation of the importance of human relations in every phase of engineering and science, then the author will feel that he has achieved one of his goals in life. You, also, will have gotten your money's worth out of the book.

Don't overlook the importance of human relations when dealing with your boss—and we all have one. Take the hints of J. M. Black: (1) Don't expect perfection in your boss—every boss has faults and virtues; (2) give him support where he needs it—know his weak points and concentrate on supporting him in these areas; (3) keep him informed; (4) learn to anticipate—try to think of solutions to problems *before* bringing them to him; and (5) be loyal—if you can't be, get another boss.[9]

Keep in mind what Dr. Robert M. Wald found [10] when studying surveys of the characteristics of company presidents. Two of the important findings were that the usual company president (1) has the ability to get along well with other people. *He is skilled in human relations, can appreciate the other person's point of view.* (2)

He sees business as people and not as buildings and machines. He has a continuing interest in the written and spoken word as a means of selling people on the idea that cooperation is desirable.

13. *Never stop learning.* You cannot keep up to date in technology or science unless you make a consistent effort to read magazines, journals, and other publications devoted to your field. Even with this regular reading you may find it necessary to take courses in new subjects. Keeping up to date is a continual problem for all engineers and scientists. To solve this problem: (1) Recognize that you *must* keep current on developments in your field; (2) determine which is the best way for you to keep current; and (3) take immediate steps to put your study program into action.

Knowledge is power. You increase your chances of making worthwhile contributions to your field when you broaden your knowledge. And, should you go into management work, your ability to achieve more through others will increase as you acquire more knowledge of people and business procedures.

So make study a part of your daily activities. Don't allow a day to pass without making an attempt to learn something new about your field. Don't let a year pass without reading at least one good book pertaining to your field. As William Feather, publisher, author, and business philosopher wrote, "Whatever your job is, there are books whose subjects cover your work. I advise every normal person to seek such books, buy them, and read them. Why spend five years gaining experience when, by the purchase of a book, you can learn what the experience of others has been?"

Education, actively pursued in your mature years, will keep your mind younger, more active. When you combine up-to-date knowledge with mature judgment and experience you are in a position to make lasting contributions to your field. Leaders in every field are those men who take time to keep their knowledge current by regular reading of key business magazines, journals, and books. They also take courses whenever they feel the need for guided instruction.

14. *Be strong—in mind and character.* Learn how to take criticism because the only time you won't be criticized is when you sit back and do nothing. Since you are now resolving to achieve more, criticism is certain to come your way. Don't wilt and give up when someone knocks your work or thoughts. Push your feelings aside, and try to learn how you can improve. Most engineers and scientists learn far more from a few words of criticism than they do from hundreds of words of praise.

Don't panic over mistakes. Find out what went wrong, and why. If you were responsible for the error, correct it. Then set up an iron-clad system to prevent the error from recurring. Tailor the system to the magnitude or importance of the error. And resolve that, "I'll never let *that* happen again." Even so simple a resolution can prevent the same error, provided you've analyzed why and how you made the error.

Printer's Ink magazine[11] said, "The man who makes wrong decisions is more in demand than the one who makes none." Five hints this magazine recommends in adjusting mistakes are: (1) Don't do anything about a mistake until you've analyzed it. (2) Consider the nature of your error, how grave it is, its effect on others. (3) Consider the causes: Was the error due to lack of facts, lack of foresight or judgment? Is it likely to recur? What can you do to be sure it won't recur? (4) It's wise to admit a mistake but don't go about it in a negative way. Through learning by your mistakes, you can build confidence in your ability to undo them. (5) When you go to admit your error to your boss, think in terms of him and the company rather than yourself.

Believe in yourself, your work, and your career. Listen to the opinion of others—then do what your mind tells you is right. Respect the opinion of others but have the strength of character to follow your own beliefs. Don't be swayed by every loud-voiced know-it-all. Study a situation; ask questions; collect information; then take the course of action you deem wisest. Trust yourself; for if you don't, how can others?

15. *Be politely aggressive—don't be afraid to take chances.* Some engineers and scientists hew to the line of meekness—they seek to constantly create the "good-guy" impression. Certainly, good guys are liked by most people. But few good guys get the big jobs because there comes a time when you must live by your beliefs. Then your good-guy pose falls apart.

Corporate survival depends on whether you're a turtle or a Tarzan, says Lon D. Barton, president of a Chicago executive-placement service.[12] Timid managers who hide their achievements and rarely stick out their necks fare badly in the competitive struggle. Conversely, the man who reminds management of what he's doing is difficult to overlook when it's time for promotion. In our cost-conscious economy, the run-of-the-mill executive is bound to lose.

How can you take the credit without making yourself obnoxious? Talk freely to the boss, but be sure what you say is important, direct, and organized. Don't waste his or your time. Initial every memo, copy, or official document. Express your ideas in simple, concise English. Suggest operating and policy changes even though you may be going out on a limb. If your idea clicks, you'll be far ahead of the man who plays it safe and mouths only company dictum. Above all, don't let others take credit for what you've done.

A positive, forceful image can also be projected in your dress. Good taste with a little dash and daring in combining colors implies your effectiveness and ability. The timid man often wears bland colors or a combination that is obviously out of place, tending to push him further into anonymity. Remember—controlled aggressiveness can pay off if you use it wisely. As Patrick B. Comer, Jr., wrote, "Getting ahead in industry is an adventure—and there aren't many people today who like adventure. . . . The supervisor who gets ahead must have a basic, materialistic, selfish interest in a better life . . . and must believe he'll reach his goal if he tries hard enough. The door is still open to initiative."

16. *See the overall picture—not just the specialized one.* Some engineers and scientists, once they are promoted to supervisory positions, have trouble seeing beyond the limits of their own little specialty. So when they are presented with problems in finance, human relations, or some other area outside their specialty, they fumble for a quick answer that is often wrong.

Every graduate engineer and scientist has, by virtue of the time spent obtaining his degree, learned how to study and master a new subject. Why, then, can't these engineers and scientists learn to see the big picture? They can, if they wish to, but many have grown so comfortable in the narrow confines of their specialty that they are afraid or reluctant to leave its reassuring atmosphere.

You can't remain a specialist if you wish to move ahead in engineering or scientific management. Whenever you are offered a promotion to a new job in which you must supervise people, you will leave some of your specialty behind. This is a fundamental fact of business life. So if you do not want to leave the comfortable confines of your specialty, you will, in general, have to decline such promotions.

Practice seeing the overall picture by studying the organizational structure of your firm. Learn the financial details of your organization by studying annual reports, stock offering circulars, budgets, and quarterly financial statements. Compare the gross business, net operating profit, and other financial details of your organization with that of other firms in the same or similar activities. Determine where your firm stands on the *Fortune* magazine list of the 500 largest corporations, or how it compares with firms on this list.

Make similar studies of your organization's sales force, service departments, manufacturing divisions, etc. Learn how your department fits into the overall structure. Then make your decisions with a view of the entire organization in mind. Having this overall view will enable you to make better decisions with fewer errors. Getting out of the rut of your specialty can be a broadening and an exhilarating experience. It can also be the first step to greater achievement and larger financial rewards.

17. *Develop a spare-time income.* As an engineer or scientist your income will be adequate for all your normal needs. But like most people, you and your family probably have interests, dreams, and responsibilities that could be partially or completely met if you had a larger income. Every time you receive a raise in salary your normal expenses seem to be a few dollars more than your new income. "How," you ask, "can a man ever catch up?" One method that has worked well for many engineers and scientists is to develop a spare-time income.

What are the advantages of having a supplementary income? There are many. The activity generating the income gives you an interest outside your daily routine that broadens your knowledge. A dollar earned at an activity other than your regular job is a discretionary dollar—you can spend it or invest it without affecting your normal way of life. Thus, you might invest your extra income in an educational fund or trust for your children. You might buy into another business, invest in real estate, etc. Or you might spend the extra income on a vacation, travel, home improvements, etc. But no matter what you do with the extra dollars you receive, you'll find that a spare-time income can enhance your life and help provide more of the things you and your family seek.

As an engineer or scientist you are in an ideal profession to earn a spare-time income. Here are ten ways that you can earn extra money without investing a large amount of capital: (1) Form a consulting firm to work in the area of your specialty; (2) write technical articles or books for publication; (3) license firms to use your patents; (4) teach in a technical institute or college in the evenings or on Saturday; (5) invest part of your regular earnings in income-producing ventures—rental real estate, a small business, the stock market, etc.; (6) work for your local election board, school board, professional society, or similar group; (7) conduct lectures or seminars for engineers, scientists, or the general public on topics of interest to them and in which you specialize; (8) convert your hobby into an income-producing venture—charter your boat, build models for a fee, take photographs of important industrial subjects, etc.; (9) review article and book manuscripts for publishers in the area of your specialty; and (10) serve as an expert witness for legal firms in cases in which you are competent to testify.

There are many other ways in which you can earn extra income. A number of these are given in various chapters in this book.

Don't ever underrate the importance of a spare-time income. Even a small extra income can mean the difference between a routine life and a comfortable one. Boost your outside earnings, and you'll derive a new sense of independence and fulfillment in your life.

18. *Build your abilities to communicate.* Learn to be persuasive in your verbal and written communications. Since most of your dealings with others are in verbal form, the more facile you become in speaking and listening, the greater your chances for outstanding success. Being able to speak well permits you to express to others your ideas and knowledge. If you fail to express these ideas often and clearly, people will not know what you think.

So begin practicing today. When talking, emphasize the positive aspects of your ideas. Human nature being as it is, many people listen to hear the negative aspect of your ideas because the negative gives them an opportunity to criticize your ideas. Certainly, helpful criticism is good—and you want as much of it as you can get. But emotionally based criticism that grows out of envy or competition is seldom helpful. So emphasize the positive—it will create a far more favorable image of yourself and your ideas.

19. *Be fair, realistic, modern, and humble.* Recognize that when people work together there are two viewpoints involved. When clashes occur, you must try to see both viewpoints in an objective way. Usually, neither viewpoint is completely right or wrong. Any solution you choose will usually lie in a broad middle ground that is the rightful property of neither viewpoint. Try to be fair at all times—people will respect you for such an attitude and will work to the limit of their capacity for you. Being fair and level-headed isn't easy, but the results are well worth every effort you make.

Be realistic about yourself and other people. See your skills, and your limitations, as clearly as you can. Don't idly inflate your opinions of yourself—someone is certain to come along and puncture your dreams. Keep the promises you make to other people. If you cannot, or will not, keep a promise, don't make it. After all, you don't like those people who break the promises they make to you.

Keep modern—don't fall behind in your field of activity. You may have to change specialties every time you are promoted—but don't be discouraged. Changing specialties, and keeping up with them, will make your mind more flexible. You will truly become an all-round engineer or scientist.

Be humble—don't allow success to distort your outlook. The higher you rise, the lonelier you become because there are fewer people on your own level with whom you can talk. So remember your place—but also remember that you were once "on the board" or "in the bullpen." Retaining your humility will help assure you that you do not return to either beginning.

20. *Keep working—your future is bright.* We engineers and scientists never had it so good. And the future promises more than we have today. Our profession is, finally, being recognized by more people. The true worth of engineering and scientific ability, spurred by the race in space, is at long last being admitted by everyone. Today as a graduate engineer or scientist you have a bright future.

Keep working. Believe in yourslf. Invest in yourself by building and broadening your skills. Work to achieve more—and watch your accomplishments grow. And as one engineer to another—good luck. May your rewards be great.

REFERENCES

[1] Auren Uris, "How to Have Time for Everything," *Dun's Review and Modern Industry,* August, 1957.

[2] Western Supply Company, "Engineering-Management Status: Dream or Mirage?" *Booklet E-9,* Heat Exchanger Division, Western Supply Company, Tulsa, Oklahoma.

[3] T. R. O'Donovan, "Your Promotion Potential," *Executives' Digest,* February, 1962.

[4] H. L. Rusch and J. R. Goeke, "What Engineers Expect from Management," *Electrical Engineering,* January, 1957.

[5] Harold E. Brewer, "The Six Steps in Problem Solving," *Supervision,* July, 1958.

[6] N.Y. Telephone Company, "You Can Add Muscle to Your Memory," *Telephone Review,* May, 1962.

[7] "Business-eye View of the Egghead," *Business Week,* Jan. 27, 1962.

[8] "The Fine Art of Creating Cooperation," *American Business,* April, 1960.

[9] J. M. Black, "Don't Make the Boss a Problem," *Supervisory Management,* May, 1962.

[10] Dr. Robert M. Wald, "What Are Company Presidents Made of?", *Public Relations Journal,* September, 1959.

[11] "How Executives Can Undo Their Mistakes," *Printers' Ink,* Sept. 11, 1959.

[12] "Is It Worth a Pain in the Neck?", *Factory Management and Maintenance,* October, 1962.

Index